# On Being a Vermonter

# On Being a Vermonter and the Rise and Fall of the Holmes Farm 1822–1923

David R. Holmes

Center for Research on Vermont
University of Vermont

White River Press
Amherst Massachusetts

First published by White River Press, Amherst, Massachusetts
whiteriverpress.com

ISBN: 978-1-887043-94-6

Book design: Emily Anderson

Cover design: Martha Hrdy

Cover image: Courtesy of Holmes family personal collection

(Permissions/photo credits not mentioned elsewhere courtesy of Holmes family personal collection)

Library of Congress Cataloging-in-Publication Data

Names: Holmes, David Robert, 1942- author.
Title: On Being a Vermonter and the Rise and Fall of the Holmes Farm 1822-1923 / David R. Holmes.
Description: Amherst, Massachusetts : White River Press, [2021] | Includes bibliographical references and index.
Identifiers: LCCN 2021012128 | ISBN 9781887043946 (paperback)
Subjects: LCSH: Farms--Vermont--History--19th century. | Farms--Vermont--History--20th century.
Classification: LCC S561.6.V5 H65 2021 | DDC 630.9743--dc23
LC record available at https://lccn.loc.gov/2021012128

*"We shall not cease from exploration*
*And the end of all our exploring*
*Will be to arrive where we started*
*And know the place for the first time."*

*—T.S. Eliot, Four Quartets*

David Holmes presents the story of his significant Vermont family, which arrived on the frontier soon after the American Revolution and in 1822 re-settled to farm in the Champlain Valley. For three generations the Holmes Farm grew with hard work, ambition, entrepreneurial skills. By late in the century the prominent farm had racing trotters and was planting apple trees for the finest orchard in Vermont. Yet, with economic hard times, the Spanish flu, and dry weather after WWI the farm failed in 1923. Once again, the family moved on, this time to Middlebury. It redirected its goals to pursuing an education for its family and 5th generation Holmes graduated from Middlebury College, as had his father. In this well-documented history, Holmes used town records, as well as clippings, family letters and photos; moreover, throughout he thoughtfully wrestled with reflections on the Vermont character and his family. Two hundred years after the founding of the farm, he has returned to Vermont to live in a home nearby on Lake Champlain.

*Travis Beal Jacobs*
*Fletcher D. Procter Professor Emeritus of American History*
*Middlebury College*

# Contents

# Preface

Life is an entanglement of family, place, and self.

This book aims to find meaning in (a) the history of the Holmes family and its farm, (b) what it means to be a Vermonter, and (c) how I have been molded by it all. I started this project forty years ago, and it was too soon. I wasn't ready—or able—to stand back and put things in context. I didn't have perspective, and I didn't know enough.

I realized too that family history in itself, even with its human interest, has limited significance. Unless one is telling the story of a great personage or a seminal moment in history, which this book does not, who really cares? What does such a narrow story add to the sum of knowledge?

I asked myself: Is the history of the Holmes farm *representative* of something larger? Does it add to our understanding of Vermont and Vermonters? With more time to think about it—forty years—I concluded that, yes, this history of one farm and one family—my family—tells us something larger about Vermont and its people. So, in the re-start of my research, I was motivated to dig deeper and to situate the story of the Holmes farm in the wider context of life in Vermont.

The reader will find that the story enhances understanding of several dimensions of Vermont history:

1. The settling of Vermont in the 1700s and the hardy pioneers who got a foothold on Vermont land and endured.

2. The emergence of Vermont agriculture, both in the foothills of the Green Mountains and the more conducive environment of the Champlain Valley.
3. How multi-generational Vermont farms emerged and sustained themselves through ever-changing circumstances.
4. How events beyond Vermont—war, the national economy, disease, inventions, migration—had an impact on the wellbeing of Vermonters and their farms.
5. The emergence of two distinctive Vermont enterprises: apple orchards and raising Morgan horses.
6. Forces that undermine Vermont farms and lead to their demise.
7. The resilience of Vermonters and how educational opportunity enables people to recover and move on.
8. The development of distinctive attributes that define a Vermonter.

It is important to emphasize that this part of the project—placing the farm's history in a larger context—was made possible by the work of others:

1. Countless Vermonter citizens who study their towns and put local history to paper.
2. Vermont town clerks who, over 250 years, have retained property deeds and Grand Lists.
3. Organizations, such as the Vermont Historical Society and the Center for Research on Vermont, with the explicit mission to advance the study and understanding of Vermont history.
4. State agencies and organizations that support and document Vermont agriculture.

5. Librarians who have collected and made available abundant resources on Vermont history.
6. Vermont journalists who write the so-called "first draft" of history.
7. Vermont historians who have made it their life's work to study, interpret and write about the state's history.

I arrived at the University of Vermont as a newly minted Ph.D. in 1974, the year that UVM's Center for Research on Vermont was created. My colleague at the university at the time (we worked together on a project at the university's Living-Learning Center), Nick Muller, was co-founder of the Center.

Forty-seven years later, Nick and his lifelong colleague in Vermont history, Kevin Graffagnino, published *Vermont Heritage: Essays on Green Mountain History, 1770-1920.*" All Vermonters owe a debt of gratitude to Nick, Kevin and the group of scholars who have made Vermont one of the best documented states in America. They have added immensely to our understanding of where we live and who we are.

Vermont is different. Vermonters are different. What follows is the evidence.

## Chapter One

# The Vermont Character

*To make sense of our world, we need to arrive at a language that captures in an understandable way who we are.*

This inquiry started with a collection of old picture albums, fraying documents, and stories passed along to me about the Holmes farm and orchard that existed in Charlotte from 1822–1923.

My aunt, who was born on the farm in 1911 and lived there until its demise, made sure that these remnants of family history stayed intact. She preserved the materials in bedroom drawers and in the attic of the house in Middlebury, where the family moved in 1923 and where she lived for seventy years until her death.

The farm ceased to exist more than a hundred years ago, but its history is important to understanding the trajectory of the family and a century-long slice of Vermont history. It is important too in understanding what it means to be a Vermonter.

When I lived in Vermont in the 1980s, I gathered up the old albums and had a professional photographer take pictures of the pictures. I saved the negatives, the contact sheet and the 5×7 prints that were made at the time. The vivid images of the farm and its people are imprinted in my mind and memory. Here were my relatives (great-grandparents, grandparents, my father, and his sister), the famous apple orchard that attracted attention throughout New England, the horses that were raised and sold widely, and the land on the shores of Lake Champlain.

At this time, I had an office near the University of Vermont library. One day I wandered over to the Special Collections room to see if there might be something about the farm inside. With the guidance of a staff member, I was directed to old annual reports of the Vermont State Horticultural Society and the Vermont Department of Agriculture. To my astonishment, the Holmes apple orchard was featured in numerous volumes during the early 1900s and contained extensive, verbatim, commentary by my great grandfather C.T. Holmes.

I copied pages from the volumes and put them in a notebook. Next, with my aunt's help, I pored over her collection of old letters and family documents. We visited the Holmes headstones in the Quaker cemetery in Monkton, where the family lived in the 1700s before moving to Charlotte in 1822.

Here was a story that I was driven to understand: The enterprise that thrived, died, and led to the family's migration. Then I stopped.

We moved away for 30 years and the idea of capturing the history of the farm faded to the background. When we returned to Vermont in 2017, I was still toting the pictures and the copied material from Special Collections. I decided to restart the search.

But my perspective evolved. I first envisioned a scholarly work that would contribute to the literature on Vermont farms and orchards in the 1800s and early 1900s. That kind of undertaking is aligned with my training and my years in the academic world. I soon recognized my emotional connection to the story, though, and realized that I could not separate myself from the objective history.

Of course, trying to both create good history *and* include one's connection to it has potential pitfalls. A respected memoirist wrote:

> Reporting on your own life or on the lives of people who share your DNA can be much more challenging than reporting on strangers. For starters, some friends and family will hate you for it. But the

> main difference is the emotional investment—and emotional payoff.[1]

So, in this undertaking, I have attempted to both recover and convey an important facet of Vermont history and reflect on the inner life of the family, my connection to it, and what it means to be a Vermonter. The heart of the project is reconstructing and explaining the history of the farm and family. The philosophical reflections flow from there.

One dimension of the story was hidden from me. When I embarked on the project, I was aware that the farm developed financial problems and fell into arrears with the bank and, because of this, the family packed up and moved to Middlebury. Looking back at conversations with those who lived at the farm until departing in 1923 (my grandparents, my aunt, and my father), there were references to the "old farm" but no talk about the event itself, the foreclosure and departure. I didn't sense deep regret or heartbreak.

After immersing myself in the life of the farm and what came after, I now believe that there was a shared silence over an event that was traumatic and heart-rending. The family's innate stoicism and resilience built up over several generations enabled them to move on, but the loss of the farm and forced departure was an impossible-to-ignore calamity. And they chose not to speak of it.

By the early 1900s, the large state-of-the-art apple orchard was known throughout Vermont and New England. The horse breeding business, built on the Morgan horse lineage, was large and of high quality. Racing the horses brought further recognition and satisfaction. The farm was a conglomerate of numerous businesses (apples, horses, bricks, wheat, sheep, cows, fruits, chickens) and was a close-knit, multi-generational enterprise that sustained the family for a century.

And then it ended.

C.T., the family patriarch and the driving force of the farm, was 66 years old when the farm failed. He and his wife, Clara, moved into the newly purchased Middlebury house of his son, Robert. They set up in the small "south room" next to the kitchen on the ground floor. C.T. died seven years later of a heart condition. Clara died soon thereafter.

The calamity was unspoken for a reason. The wound was deep, permanent and sad, and their psychological survival required that it be silenced. To seal away the hurt, the family went quiet about what happened and, at the same time, reoriented themselves to creating a future in Middlebury. C.T. never recovered from the loss but the others—his son and daughter and their families—created different and successful lives away from farming. They started over and, as a clan, began life anew.

## A. Claiming Vermont Heritage

According to crusty old timers, there is a simple, no-nonsense definition of a Vermonter: You have to be born here. Therefore, I am a Vermonter, just barely.

Lucky for me, I was born in Rutland Hospital in March 1942. But I didn't stay long. Four months later my father came back from Quantico, Virginia, as a newly anointed Special Agent of the Federal Bureau of Investigation, and gathered up my mother and me. We boarded a train and headed to Texas for my father's first posting with J. Edgar Hoover's wartime FBI.

Born in Vermont, you are always a Vermonter, except for one disqualifier—do not move away and return with an air of self-importance.

When a native Vermonter senses an air of superiority, there is an easy-to-miss but pointed response. You may get an askance look from the cashier at the local grocery, a curt reply from the town clerk to a question, or an eyeroll from the carpenter working at your home. If so, you are outed and you don't even know it. Of course, a worse situation is having no Vermont roots, moving to Vermont from "away" and bringing attitude. The grizzled old guy at the town garage will peg you with a pithy epithet and that is that.

As a Vermonter by birthright, I moved back to the state three times: Four college years in Middlebury, a job for 13 years at the state university, and finally one last move home at age 75 in 2017. By that time, I had been absent for 58 years. In T.S. Eliot's eloquent and oft-repeated stanza, I did not "cease from exploration" during these years away. During this time, however, I knew in my bones that I was just visiting.

In the years before returning, my wife and I decided it was time to come home. Each of our stops was meaningful, but it wasn't Vermont and we missed it. We missed seeing our relatives, the ease in getting around on roads and highways (i.e., the lack of bumper-to-bumper traffic, except for Shelburne Road at rush hour), the four defined seasons, the moisture (13 years in semi-arid Idaho were enough), and a Vermonter's undercurrent of humor. This is what we remembered when we loaded up a U-Haul trailer in Idaho in 2017 and drove east with the dog. We moved to Panton on the shores of Lake Champlain, arriving in the state where we started.

But what did we know? We left in 1987. Had Vermont changed in fundamental ways over that 30 years? Was there more than one Vermont, an urban version and a rural one? Did we romanticize what Vermont is all about?

So, late in life, a personal, existential, question loomed: Am I, on returning home, an authentic Vermonter? Will I arrive where I started and know the place—and myself—for the first time?

## *Defining a Vermonter: The Evidence*

Since the state's founding, locals and outsiders have ventured thoughts about what it means to be a Vermonter. An attempt to capture the essence of Vermonters, however, has inherent obstacles.

First, whatever idealized attributes we may attribute to Vermonters, they—being human—do not display them consistently. Moreover, a generalization is just that, a statement that obscures a complex reality. Yet no one is perfectly consistent in who they are and what they do, so capturing their overall character—their best self—is good enough.

Also, we need to remember that generalizations about a place and a people have meaning for those living today and foster a sense of community. A common understanding about who we are as Vermonters has communal value.

Second, many Vermonters live in cities that may have a different culture and identity than their neighbors tucked away in small towns or the backcountry. However, Vermont cities are relatively small, and many city dwellers come from small communities and bring attributes of small-town Vermont. Many of those who come from out-of-state urban environments make the move because they are seeking Vermont ways of life, and they quickly take to Vermont and all the state has to offer.

Third, perhaps Vermont has changed so profoundly that connecting the past and present—finding meaning in the span of Vermont's history—is impossible. The changes include the revolution in communication technologies, the amalgamation and growth of health care systems, the appearance of national chains of box stores, out-of-staters buying up land, the vastly increased size of farms (especially the financially viable ones), and many others that Vermonters can enumerate. Yet Vermont is still a state on a small scale with people who want to keep it this way. Politics still has a local dynamic. We know each other. Most importantly, the core values are resistant to change.

Fourth, in the end, maybe the attempt to capture the essence of a people is a fool's errand. But many scholars before me have had the audacity to attempt to describe a people. I have several works on my shelf that succeed in this aim, including: St. Jean de Crévecceur's letters in the 1700s that address "What is an American;" *The American Mind: An Interpretation of American Thought and Character Since the 1880s* by Henry Commager; *The Lonely Crowd: A Study of the Changing American Character* by David Riesman, Nathan Glazer and Reuel Denney.[2]

In setting out to define Vermonters I have attempted the same thing on a smaller scale.

These objections notwithstanding, I conclude that figuring out the Vermont identity is a legitimate undertaking. To this end, I turned to three kinds of evidence: What informed observers have said; the legacy and lessons of the Holmes farm over its 101-year history; and reflections on my own experience.

At the outset, let's agree that the "you have to born here" standard for being a Vermonter is inherently misleading and inaccurate. First, during the colonial era, the "first Vermonters" came from elsewhere: Connecticut, New York, Massachusetts. Second, there are many hundreds of Vermont residents, born elsewhere, who are truer Vermonters than some who were born here and have lost their "Vermontness."

As Graham Newell, who served as a state legislator for 26 years and was a seventh generation Vermonter, said: "I'm not one of those who says that you've got to be born here to be a Vermonter. If you are a Vermonter, you feel like one and you don't have to explain it."[3]

## *Expert Opinion*

Observers have sought to capture Vermonters and their character since the Revolutionary era. Early on, Vermonters tended to define themselves by what they were *not*: "We aren't like those prissy New Yorkers down by Albany who want to take us over;" "We aren't royalists like Governor Wentworth on the New Hampshire side of the Connecticut River who thinks his land grants will determine our future." The early settlers were tough-minded, often ornery, independent farmer-fighters who loved the land. *Their* land. And they got their way when Congress made them a state in 1791.

Because the early settlers were so busy getting the British off their backs and establishing farms and communities, the indoor sport of describing a Vermonter came later.

So, jumping ahead in time, what can we learn from those who have contributed to the discussion about being a Vermonter? In *Vermont Tradition: The Biography of an Outlook on* Life, Dorothy Canfield Fisher aimed to see how Vermont history shaped, molded, and created Vermont character.[4] Her 1953 book surveys Vermont's history and contains countless astute observations. According to Ida Washington in her essay on Fisher, Fisher's most telling observations about the Vermont character are its strong individualism and old-fashioned grit:

> A belief that every person has a right to live his own life in his own way, with only the most elemental requirement of cooperation with the community, and its corollary, a respect for individual human beings based on qualities of character independent of wealth, titles, or other external matters.

> An ability to persist and endure in basic convictions in the face of personal problems and economic and social changes as sudden and unpredictable as Vermont weather.[5]

Writing in 1990, William Mares, whose works include the provocative *Real Vermonters Don't Milk Goats* (with Frank Bryan), added additional propositions. He posited four long-standing Vermont values: Vermont is egalitarian; Vermonters have a strong sense of community; Vermonters' desire to control one's own life; Vermonters value a working landscape with working farms and productive forests.[6]

Mares differed with Fisher on the degree to which community is a driving motivation, and I believe he had it right. From all evidence, community is a fundamental element of Vermont life.

We have such things as town meetings ("Property taxes are killing us!"), Front Porch Forum ("I have leftover plywood for anyone to pick up."), local action on divisive issues ("Don't close our small school!"), and joining together to build a hiking trail ("Be at the Chipman Hill trailhead at 8 a.m. on Saturday morning."). In 2021, if you read your weekly newspaper or saunter through town with your ears open, you find this close-to-home chatter.

This is what Vermonters mean by community.

Mares cited developments impinging on these core values and suggested that there is a "schizophrenia" in the Vermont state of mind. He wrote:

> It is surely correct to say Vermont and Vermonters are riven with contradictions. I see impulses to join in America and to run away; to build four-megabit microchips and heat with woodchips; to have an integrated fiber-optic network and still talk baseball and weather at the post office; to provide for all but preserve the neighborhood from the homeless, trailers, and condominiums.[7]

Some 30 years after Mares wrote his essay, I find that Vermonters have learned to live—to coexist—with these contradictions. This is evidence, I believe, of a healthy New England pragmatism.

Another keen student of Vermont history, Gregory Sanford, reinforced Mares point about the impact of change:

> An unchanging, idealized identity masks changing realities. When we can no longer ignore the new realities, we are confused and angered by the discrepancy between our self-portrait, idealized and unchanging, and our actual life style.[8]

Despite the lack of consensus and an occasional naysayer about attempting to capture the essence of being a Vermonter, the effort is unlikely to cease. To make sense of our world, we need to arrive at a language that captures who we are in an understandable way.

A 2009 document at the Sheldon Museum of Vermont History in Middlebury posed some fundamental questions that are pertinent to this inquiry. The authors turned to the eloquence of Robert Frost ("I took the Road less traveled by, and that has made all the difference") to frame questions about being a Vermonter:

> Vermont has long been a place with a strong personality that also expresses a larger truth about his adopted state: rural, crusty and independent. Over two centuries, Vermonters have continued to live in ways that have become obsolete in other parts of America. We still dwell in towns, villages and farms rather than large urban areas. We practice democracy at town meetings and vote with penciled Xs. Most of us know our neighbors. Our people and our landscape are known for being special. We are proud to have taken the road less traveled, whether through choice or through necessity.
>
> What historical forces have tended to differentiate Vermont, and Vermonters, from the

> mainstream? How are those differences reflected in this distinctive landscape, where so much sense about ourselves resides? In what ways did our experiences reflect broader national trends? How have we taken "the road less traveled" and has it made all the difference?[9]

## *Legacy of the Holmes Farm*

Although we have adequate documentation of the Holmes family history for the late 1700s and the early 1800s (the coming to Vermont and creating homesteads), this story focuses primarily on the 101 years of the family farm and orchard, 1822–1923. There are ample records from this period and, equally important, I knew relatives who lived on the farm—my grandparents, my father, and his sister—and other relatives who spent time there. My proximity to the farm's history and its people helped me interpret its culture and dynamics.

The story of the Holmes farm is a singular contribution to Vermont history. A survey of sources on Vermont agriculture indicates that there are no case studies of a Vermont farm that span the early 1800s to the early 1900s. There are a few studies of farms of more recent vintage, but the earlier times lack in-depth examination.

This history provides unique insight into the culture, lives, economics, and fate of a farm that existed during the state's formative years.

This story is one of a kind, but the Holmes history is not unique. Like many before and after, the family sought opportunity by moving on. They came from England to Connecticut in the 1660s, went north into the Vermont wilderness in the late 1700s, and established a farm on the west side of the Green Mountains in the town of Monkton, which lies 30 miles southeast of the present city of Burlington and extends along a ridge line with farmland falling off to the east and west.

In 1822, Nicholas Holmes acquired a more promising property on the shores of Lake Champlain northwest of Monkton in the town of Charlotte. The new property was larger, flatter, and more conducive to large-scale plantings. Lake Champlain, which stretches north into Canada and south toward Albany and the Hudson River, enabled the farm to transport its products more easily.

Sparked by this move, the Holmes farm grew in scope and reputation, highlighted by a flourishing apple orchard, the breeding and racing of trotting horses of Morgan heritage, a brickyard, and a family presence in Charlotte and Vermont.

Like almost all farms, however, the Holmes farm experienced adversity, including bad weather, illness, poor crops, a volatile economy, land disputes, and accidents. In the end the economics were too much. A shortfall in apple sales and a substantial loan that could not be repaid resulted in the farm floundering between 1915–1922. The Holmes family fell in arrears on scheduled mortgage payments and, finally, in April 1923, creditors foreclosed on the farm by calling in the loan.

In the face of bankruptcy and no prospect of getting clear of debt, C. T. Holmes and his son Robert called it quits. They moved 25 miles away to Middlebury, a vibrant small town and home to a liberal arts college, to start over.

The Holmes family farm was no more.

This book describes a defining chapter in the history of a Vermont family and their farm along Lake Champlain that lasted for a century. At the end, the center did not hold. The farm failed and the branches of the family moved on. The epilogue of the story is the diaspora of the Holmes family—where they went, what they did, and how "Vermont" remained in their soul.

The fate of the Holmes farm was not a unique event. The history of Vermont is replete with stories of farms that failed to endure. Dealing with adversity is a condition that many thousands of Vermonters have had to confront. Out of necessity, they moved on and reoriented their lives.

### *My Journey*

I was born in Vermont 19 years after the family left the farm. I returned later to attend Middlebury College and worked at the University of Vermont for 13 years from 1974–1987, but I lived elsewhere for most of my life. My exposure to Vermont and its essence has been intermittent and shaped by the meandering trajectory of my life. Two experiences stand out: I bought a Vermont farm and I married a Vermonter.

I was influenced—at least subconsciously—by my family's agrarian history when I took a radical step. On a trip back to Vermont from Illinois in 1969, I bought a run-down, non-operating farm of 150 acres in Moretown on the east side of the Green Mountains. Jones Brook Road was at the foot of a mountain, and I soon discovered that it washed out every spring. The property cost $21,000, and I had a ten-year note.

A few years later, a college friend and I bought 50 acres at the top of the mountain behind the farm, Chase Mountain, for $1,800. The summit looks across the Mad River Valley, and you can see the Sugarbush ski area when the leaves are down. We had it logged twice over the years. It is hard to get to, and local hunters consider it their personal property because it is prime deer country. If we ever posted it, we would be asking for trouble and, besides, who would enforce the prohibition? We have paid annual property taxes for more than 40 years, and it still sits there for all to use.

I hold unproductive land at the top of a mountain in Vermont, and that's okay.

I never stayed long enough at the Moretown farm to raise any crops or fix up the decaying farmhouse, but I created a lawn around the house and camped out whenever I visited. I met the neighbors, most of whom were barely hanging on financially. I was shown the immense gun collection on the walls of Leo Ciampi's small house and had coffee with Leo and his wife, Dot. Their 12-year-old son, Mario, was already driving every machine on the property—snowmobile, ATV, tractor, the old car—and hunting with Leo.

Several years later I was called out of a class I was teaching at the university in Burlington to take a phone call. A Vermont State Police officer informed me that the farmhouse was on fire and likely a total loss. Someone had torched it. Eventually, I sold the property but, for a few years, it had been my land. There was something meaningful, even sublime, about being there and having those 150 acres. It was a part of Vermont—the original Vermont—that had been left behind.

My neighbors were authentic Vermonters. They scratched out an existence, hunted, fished, cut wood, and told stories about events on Jones Brook Road. One story told to me by a neighbor hit close to home: A previous owner of my farm had shot his wife while she was sleeping in bed. Murder in the neighborhood! In fact, several neighbors liked to say "There ain't no law on Jones Brook Road." No matter—they loved their lives, and I loved them.

I married a story-telling Vermonter in 1976. Fortunately for my wife, the definition of a Vermonter has some flexibility. She was born across the Connecticut River in Littleton, New Hampshire, and moved to Vermont with her parents when she was three.

Narrowly defined, this doesn't make her a Vermonter. Still, she has the traits of one. In addition to an immense work ethic—with a heavy dose of physical labor—that compares favorably with any Vermont farmer, she finds amusement in everyday human interactions around town. In addition to her sharp memory, her Vermont anecdotes are embellished with a dead-on Vermont accent. Among countless vignettes, here are prototypical examples.

> Caleb and I were driving on Route 103 through Chester on the way to Middlebury. We slowed as we approached a diner on the outside of town. Everyone in the diner had come outside to consider the problem of a tractor trailer stuck on a narrow bridge going through town. When I asked the owner of the diner if there was another way around Chester, she replied without missing a beat, "How much time do you have?"
>
> I stopped at a gas station in North Ferrisburgh to buy a newspaper. The woman behind the deli counter was talking about being stopped for speeding on Greenbush Road. I commented that I had recently been stopped there too. One woman said, "It's rigged. I wasn't going any 51 miles an hour when I was stopped." She asked me if the

> sheriff who stopped me "was it a little guy with a mustache?" I said, "No, it was large woman with blond hair." The other deli lady cut in to say "That's Brittany Trudo! Brittany would give you a ticket for low air pressure in your tires."

To get the full effect, you had to be there for the story, accent included. But what makes these stories amusing? And what does it tell us about the unique brand of Vermont humor?

- The situations are ***familiar*** (traffic snarl; rural police giving tickets on an open road).
- The person with the punchline takes a subtle dig (toward the lady with out-of-state plates; toward the over-zealous cop who is probably a neighbor) but is ***not mean-spirited*** about it.
- The diner owner and the deli lady see life ***beyond the purely literal*** (i.e., they have a creative intelligence that sees life at multiple levels).
- They are having fun in poking fun at the ***small absurdities of life***.
- The humor is low-key, and you *must* ***listen*** **to get it**.

This is not knee-slapping humor, but it captures a dimension of how Vermonters see life. Her stories reveal an undercurrent of everyday humor and how Vermonters, in a good-natured way, poke fun at others and themselves. Fortunately for my wife, a collector of humorous asides from the mouths of Vermonters, almost no one knows she came three years late to the west side of the Connecticut River, so she gets a pass.

I have found this to be true: Vermonters are fundamentally good people who love where they live (e.g., the Ciampi's on Jones Brook Road). And Vermonters, if you listen, have humorous stories that reveal tacit truths about themselves. These experiences became part of my education as a Vermonter.

## B. Family History: Meaning and Method

Numerous archives of the Holmes family have been saved and preserved. There are hundreds of old pictures collected in albums or piled in boxes. There are family trees, neatly rendered, created by relatives. There are old school diplomas, ancient cookbooks, expense ledgers from the farms, letters, and fragments from diaries. There are descriptions included in local histories, agricultural journals, and editions of The Colonial Genealogist.

As rich as this mother lode of information may be, these are fragments—random data points—of the past. They lack a narrative, a story, a theme. In their present form they lack meaning.

The first task is to search for meaning, if it can be found, in an archival collection like this. Otherwise, these fragments will be just that for an eternity. This search, for clues and insight about one's family and self, is where the excitement lies.

It may be that the world has moved on in such a way that this century of family and Vermont history, 1822–1923, has no real interest or relevance to those living in the present: Parents, children and grandchildren of the extended Holmes family; citizens of Charlotte, Monkton, Huntington, and other towns connected to the story; historians of Vermont's past.

In stark contrast to the life experience of those living 200 years ago, we live in a world of high-speed internet, encompassing social media, and instant connectivity; jobs that are transitory (the concept of a single lifelong career has disappeared); educational attainment that assumes a college degree or beyond; families that are separated by geography (our children go where the jobs—or ski slopes—are); gender roles that are rapidly changing; a society that shares a concern about stress and addiction.

Nicholas or Jonathan or C.T. Holmes would be mystified, perhaps shocked, by all this.

## *Importance of Family History*

Except as a source of curiosity ("The man with the long beard in that old oil painting is really my great, great grandfather? That's weird!") or the compulsion of an aging relative ("Well, at least it gives—*fill in a name*—something to do in his final years."), why care about family history? Why dig into genealogy? Why care about finding meaning in the past? For me, it was a personal quest. But looking more broadly, there are compelling reasons for translating the past for those living today and in the future.

### *Mental health*

There are legitimate concerns about the mental health of our young people. Levels of adolescent depression and suicide are at all-time highs. Young people are immersed in and shaped by social media and its dangers, pressures to compete successfully in school, and, like their parents, feel anxiety about jobs and money.

All too often our youth are atomized, anxious human beings, floating in space, who do not know from whence they came or where they are headed. The COVID-19 pandemic exacerbated the situation.

A modest antidote is to know one's family roots, to know the people and history that preceded them. This knowledge helps solidify one's identity and grounds a person in time. This knowledge is insulation against the rootlessness of social media, the disorientation that comes from "social distancing," and ongoing cultural and economic change.

### *Guidance and inspiration*

Studied closely, the past can provide guidance and, on occasion, inspiration. C.T. Holmes took a calculated risk by borrowing substantial funds to purchase irrigation equipment for his apple orchards. He was one of the first orchardists in America to irrigate in a systematic way, but the move ultimately contributed to bankruptcy. The lesson may be "don't bet the farm" when external forces—weather, market forces, inflexible bankers—may do you harm. Yet, the story may be inspiring. C.T. had a large vision and became an industry leader. Youth of a future generation might say, "Perhaps I can go after big ideas too."

*Spiritual connection*

There is a spiritual element. For most of us, when a grandparent or parent is laid to rest, that's pretty much the end of it. The sorrow ebbs over time, and the dead are mostly forgotten. They are rarely spoken of, and they are thought of only occasionally. Forgetting, as unintended as it may be, dishonors those who made us possible and shaped who we are. Resurrecting the past through family history is a powerful way to honor our relatives. Capturing their essence—flaws and all—is something they deserve. Remembrance of things past is our final thanks.

*Building a sense of community*

Family histories produce insights and lessons that have relevance to other people and families. They extend beyond one's self at a time when our world is vibrating with change and uncertainty. Usually, family history is also **local** history. Family histories, particularly when several have a common geography, create a foundation of insight and understanding about the community where we live. This brings us together as a people and creates a foundation of common understanding and sense of community at a time when so many forces pull us apart.

*Self-understanding*

Finally, no matter how far back in the past, we are genetic products of those relatives who came before. To know something of our relatives is to know something about who we are: our personality, our physical attributes, our passions, our foibles, our aspirations.

These attributes do not determine who we are or will become, but such awareness is a part of self-understanding. For me, the search for self-understanding—for what it means to be a Vermonter—is why I embarked on this inquiry.

The picture below (Figure 1.1), taken at the Holmes farm, circa 1911, shows Robert, Hannah, C.T. (from left, my grandfather, great-grandmother, great-grandfather), and visiting cousins. The image reveals something that I saw in the members of my family—a zest for life, a love of physical labor, and the pleasure of taking a break for mid-day dinner. It feels familiar, as though I could step into the picture.

*Figure 1.1 Family at noon*

## *Searching for Clues*

The Holmes farm came into being 200 years ago when Nicholas Holmes purchased 215 acres along the shores of Lake Champlain. There is useful archival material to draw on in reconstructing the history of the farm. Yet, the historical record has large gaps, especially for the 1822–1885 period.

For these years, there are no letters, no photographs, and limited entries in town histories. The most reliable documentation resides in the Charlotte Town Clerk's office, where every land transaction and every annual Grand List since the 1700s has been saved in old volumes.

Unfortunately, many of the oldest documents are incomplete or impossible to decipher. They reveal little about the triumphs, challenges, and daily life of the early years of the farm. Of course, one can infer insights from what is known generally about an historical era. We know from later records, for example, that apple orchards were not an important factor in the Vermont economy—or the Holmes farm—in the first half of the 19th century.

Yet, all in all, the early record is hazy and unsatisfying.

Tracing the Holmes family history is also confounded by the frequent use of two names: Jonathan (John) and Nicholas. Over just three generations of direct lineage, from 1716 to 1820, there were five named "Jonathan" (or "John") and four named "Nicholas." As evident in the Grand Lists of Charlotte, Jonathan was often shortened to John. It takes careful reading and rereading of the genealogies to keep the lineage straight. Also, Jonathan's son, Charles Titus, the driving force behind the apple orchard, is referred to his commonly used appellation, "C.T." In the following pages, I distinguish names and people as clearly as possible.

The history of the farm is much more robust and revealing in the later years. Even here, however, we lack a deep reservoir of source material. Thus, what follows relies extensively on the history of the apple orchard captured in the journals of the Vermont State Horticultural Society and the Vermont Department of Agriculture, letters that provide insight into life on the farm, photographs from around 1900, deeds that document the evolution of the farm until its foreclosure in 1923, and general histories of Vermont, Charlotte, and the surrounding communities.

The photographs and letters have a particular importance. The black and white images from the early 1900s bring the farm and its people to life in a way that old documents and a narrative cannot. Yet, one must remember that these black and white pictures, often faded from age, miss the exquisite days of sun and blue skies. Vermont's weather is one of extremes, and there are countless pristine days that bring brilliant blue skies, bright green vegetation, and glimmering lakes.

Dorothy Canfield Fisher, mentioned above, wrote perceptively about the ways of Vermonters. Although her reputation has suffered because of her involvement with Vermont's eugenics movement, she had a fine grasp of her adopted state. On the value of letters among family members who lived apart, she wrote:

> It is only now that we understand how these streams of letters incessantly coming and going made a great web of intimate personal relationships, living in letter instead of in face-to-face meetings. The reason why their letters are now so interesting to us is because they were conversations-on-paper, concerned as family talk always is, with the events of everyday life. It was no tremendous event for these ancestors of ours to write a letter. Their pens were practiced, they wrote the news of the day or this week to their families, as they would have called from one house to the next if they had lived next door.[10]

The letters of Holmes relatives capture a time, place, and mood in a way that a narrative summary by me cannot, and I have included lengthy excerpts from the letters I have recovered.

It is noteworthy, perhaps, that an autographed first edition of Canfield Fisher's book has been handed down over 70 years within the family. Remembering discussions with my grandparents and other relatives, it is evident that the themes of the book resonated with the Holmes family.

The articles about the orchard that appeared in state publications are quoted at length. In fact, as pointed out above, some articles include *verbatim* transcriptions of presentations made by C.T. Holmes. What a gift! Through them we can almost hear how C.T. spoke and thought, more than a hundred years later.

Unexpected discoveries and finding material after a long search are an exhilarating part of research. In this project, discovering old letters from the 1880s, finding C.T. Holmes's entries in annual reports of the Vermont State Horticultural Society, and hearing from a former resident of the property about how his family dug up and dumped the decaying trees into the lake added greatly to my understanding.

There is one regret, however. I began the serious phase of the project 35 years after I started looking into old albums, learning about the orchard, and gleaning pieces of family history from my aunt. In the 1980s she showed me the pictures, took me to the Quaker cemetery, and provided glimpses of life at the farm. She died in 1993. When I restarted my research in 2018, I was too late to ask her questions. Countless times I said to myself, "Well, I will ask Marion," and I instantly knew the impossibility of that.

Stepping back, perhaps what is most important in all of this is capturing the overall trajectory of family history, from arrival on the frontier, to building a thriving farm on the Lake, to an abrupt ending and the dispersion of the family to other places. It is family story and a Vermont story, not unlike the stories of hundreds of other families throughout the history of Vermont.

Chapter Two

# Paths to the Lake

*They purchased lots sight unseen, at best with a vague impression from speculators, and packed up and moved north.*

This history tells how three families of the northern frontier—the Holmes, Johns and Ross families—came together by marriage in northern Vermont.

Vermont in the late 1700s and early 1800s was a landscape of deep forests, crude roads, abundant wildlife, and small vulnerable farms. With the American Revolution just ended and the War of 1812 still to come, America itself was still a fragile entity. Soon, though, a stream of pioneers headed north from southern New England. Historian Edwin Rozwenc wrote the following:

> Thousands of new settlers were attracted to Vermont after the end of the American Revolution. In the four decades from 1790–1830, Vermont gained in population 79,000, 63,00, 18,000, and 44,000, respectively. This migration into Vermont, however, was a part of the normal process of expansion which had been going on in New England since the middle of the seventeenth century.[1]

Those who settled in western Vermont—the Lake Champlain side of the Green Mountains—came primarily from Connecticut and eastern New York State. These newly-arrived Vermonters knew instinctively that owning land, farming it, and creating communities was insurance against the vagaries of national politics and the continuing British presence on the continent.

This flow north included the Holmes family, who, in 1778, came to Monkton, a hilly town of jagged angles midway between the Green Mountains and Lake Champlain. Two years earlier, in 1776, the Johns family came up the Onion River (the Winooski) to Huntington, which is at the base of the Green Mountain range.

Shortly thereafter, the Rosses also came to Huntington. As one Vermont historian pointed out:

> There was land and plenty of it in the Green Mountain country. Everyone in southern New England knew this, but they were poorly acquainted with its quality and topography. In the rush of speculators to buy lots little attention was paid to location. As few of the original prospectors ever visited their holdings, lands were disposed of in cheerful ignorance of locality or worth.[2]

This passage captures the approach of the Holmes, Johns and Ross settlers. They purchased lots sight unseen, at best with vague descriptions from speculators, and packed up and went north.

Were Nicholas Holmes and Jehiel Johns, patriarchs of two sides of the family, surprised or disheartened by what they found when they arrived? Not likely. Instinctively, they put their heads down and made it work for their families. Moreover, Connecticut and New York, from whence they came, had a hilly, rocky geography that, with effort and ingenuity, had been civilized.

What could be so different about Vermont?

So, tucked away in communities like Huntington and Monkton, the Holmes family—and its future relatives by marriage—were early settlers. These pioneers labored to create a secure foundation for their families. First and foremost, they needed to establish themselves financially, and that meant starting up a farm.

Many strands of the family converged after 1822 at the Holmes farm. To grasp the essence of the family—its personality and capabilities, its victories and defeats—one must look to these early arrivals on the Vermont frontier.

## A. The Holmes Family Comes to Monkton

**Holmes line of descendants from England to Charlotte, Vermont, and Lake Champlain**

1. Francis—from England to Stamford, Connecticut before 1643
2. John (born 1630) from England to Connecticut with his father, Francis; moved to Stamford, CT, then to Bedford, CT
3. Jonathan (born 1674) moved to Greenwich, CT.
4. Jonathan (born 1716) moved to Nine Partners, NY
5. Nicholas (born 1742) from Nine Partners to Saratoga, NY, and then to Monkton, VT, in 1788 with wife and two sons
6. Nicholas (born 1782) from Monkton to Charlotte on Lake Champlain in 1822 with his family

Considering that the Plymouth Bay Colony was formed in 1630, the Holmes family left England among the early wave of settlers in the New World. No available records exist to indicate the driving motivation for the decision to leave home, whether religious, financial, or family dynamics, but, since the family identified and worshiped as Quakers in the New World, religion was a likely factor.

Here is the trajectory of heads of family from England to the shores of Lake Champlain:

1. The first Holmes to arrive on the shores was **Francis**, who left Beverly, Yorkshire, and arrived in Fairfield, Connecticut, in 1643. He brought his wife, Ann, his oldest son, John, and three other children. Francis, a blacksmith by trade, died in 1671.
2. **John** was married in 1659 and had 10 children, including Jonathan who was born in 1674. He moved the family to Stamford, then to Bedford, Connecticut, and gave land to each of his three sons, including Jonathan.
3. **Jonathan**, married in 1707 and had 7 children, including another Jonathan.
4. The **second Jonathan, Jonathan's son,** was born in 1716 and lived in Greenwich and Nine Partners, New York. Nine Partners was a staunch Quaker community that sent many brethren to Vermont to start numerous Quaker communities. Jonathan's first child, Nicholas, was born in 1740 in Nine Partners and died at birth. The second Nicholas was born in 1742.
5. The surviving **Nicholas** married Phoebe Titus and lived in Nine Partners. They had 7 children. With the end of the Revolutionary War and the weakening—but not eradication—of the British presence in the Lake Champlain region, many in southern New England looked to Vermont for land and economic opportunity. Phoebe Titus' family bought land in Vermont and, on Feb. 5, 1778, Nicholas acquired from Phoebe's father, Samuel, and John Titus "one whole shear or right of land in the Township of Monkton, State of Vermont."[3] Ten years later on March 3, 1788, Nicholas and his sons, Jonathan (1777–1849) and Nicholas (1782–1863), departed for Saratoga, New York, and then

proceeded on to Monkton. Phoebe and the other children arrived later the same year.

6. The younger **Nicholas** worked alongside his father (the elder Nicholas) and his older brother, Jonathan, on the Monkton farm. In 1822, however, he broke away from the family settlement and purchased land 20 miles west on the shores of Lake Champlain in Charlotte. Soon thereafter he brought his wife, Sarah, and children to the property. Nicholas's youngest child, Jonathan, who succeeded Nicholas as proprietor of the farm, was two years old at the time. This move was the origin of the "new Holmes farm," culminating a family journey of 179 years from Beverly, England, to Vermont's Champlain Valley.

Traveling north and making a home was not an easy thing. Christopher Wren has written insightfully about what it took:

> The journey was best taken in late winter, when the trammeled snow stayed frozen enough to make travel easier, before ice thawed into a springtime morass of mud. Families who could afford them rode sleighs or pushed sleds up the frozen Ct River or Lake Champlain. They herded cattle and sheep through valleys and foothills, passing marginal tracts in expectation of better land ahead. Wives and children walked if there was no room in the oxcarts or sledges stuffed with farm and household implements.
>
> The more intrepid sallied forth to claim and clear their land, building a hovel habitable enough for families to join them after the next winter. The earliest homes in the New Hampshire grants looked nothing like the tidy farms left behind in Connecticut and Massachusetts. Logs were felled and notched to frame a cabin. More logs, split and

> hewed, made a floor. Bark covered the roof. Sticks plastered with clay became walls. A large rock flanked by smaller stones formed a fireplace. The huts filled with smoke and leaked with rain, and were occupied before they were habitable. In winter hefty logs would be fed into the fire and kept ablaze, not just for warmth but also to keep away predators. Homesteaders slept on their unburned woodpiles, swaddled in animal pelts.[4]

This was the destiny of Nicholas and his two sons in early 1778. Although Monkton was already a defined settlement, this was still a primitive time with a few families spread over many square miles.

Monkton was a part of the New Hampshire grants, chartered by Gov. Benning Wentworth in 1762. The Holmes family arrived in a second wave of Monkton settlers.

The 1800 U.S. Census reported a Monkton population of 880, including 331 children under the age of 10. Monkton families were making babies at a fast clip, which in time supported the work of the family farm. In 1800 Nicholas was listed as the head of family of five.[5]

Nicholas farmed in Monkton for 34 years until his death in 1811, with his children linked by marriage to Vermont families such as Mabbett, Thorne, Wing, Rogers, Hazard, and Smith. Nicholas was considered a large landowner in Monkton and also owned land in nearby Ferrisburgh, where he apparently bought 100 acres from the family of Vermont's foremost folklore writer, Rowland Robinson.

### *The Quaker Heritage*

After coming to Monkton, the family became active in the Quaker community. Nicholas' oldest son, Jonathan, played a prominent role in buying land to build the first Quaker church for the Monkton Quakers. Along with Nathan Hoag, of Charlotte, Jonathan purchased 1.5 acres on behalf of the Monkton Friends for the construction of a church in Ferrisburgh. As it turned out, other Monkton Quakers changed their mind about building in an adjacent town, and both towns ended up with a church.[6]

The Holmeses were also partial to Monkton, and they attended the town's first Quaker church in the northwest part of town. The Quaker cemetery in Monkton was established on a hill behind the church with a view southeast into a valley. The remains of Nicholas (1742–1814) and his son, Jonathan (1777–1849), are buried there along with other relatives. Nicholas's' headstone (Figure 2.1) sits at the left of the family row. The rolling hills of Monkton are in the background.

*Figure 2.1 Quaker cemetery in Monkton © Emily Anderson*

In his flowery language, William Higbee, a local historian who visited the Friend's burying ground in 1899, wrote the following:

> It carries the man of today a long way into the past to walk among the old graves. many of them unmarked, and read the names. It was not the custom of the society in those days to do much adorning, either in life or death. Mankind finds here its final resting place. The quietness and the decorum of the living, passing from mound to mound evinces a desire to disturb not their dreamless sleep.[7]

To understand something of the Holmes family is to understand the Quaker community and culture. Rowland Robinson's family created their family homestead in Ferrisburgh, which exists today as a museum on Route 7. His father was a committed abolitionist, and the homestead was a station on the Underground Railroad.

Looking back in 1901 to his childhood, Robinson's *Recollections of a Quaker Boy* described the meeting house in Ferrisburgh before the years of the Civil War:

> The broad-brimmed hat, the shad bellied coat, with its narrow standing collar, the drab sugar-scoop bonnet, the scant sleeves and skirted gown with white kerchief across the bosom, the addressing of every person by the singular pronoun, the naming of the months and the days of the week by their numbers, seemed not so strange to childish eyes and ears as did the dress and speech of the 'world's people.[8]

This description, especially plainness of dress and lack of ostentation, captures the Quaker traditions and the "testimony of simplicity" brought by the Holmeses from Nine Partners to Monkton. Eventually, Nicholas Holmes left Monkton and its Quaker meeting house for Lake Champlain in Charlotte. Yet, for many years, members of the family continued in the Quaker tradition by traveling to the Ferrisburgh meeting house. In time, however, most members of the family gravitated to the Congregational Church in Charlotte.

## *Slavery, Abolition, and the Presence of Black People*

Looking back to Nicholas and Phoebe's time in Nine Partners, New York, a sobering entry in the Quaker records of Nine Partners was the following: "Manumission of Slaves, 4–4–1776, Nicholas Holmes and wife Phoebe, Negro boy, Nathaniel, aged about 8, when 21."[9]

We don't know what granting freedom, apparently at age 21 in 1789, to an 8-year-old boy really meant. Did Nathaniel remain with the Holmeses? Did he join his own family someplace else? Knowing there was substantial slave ownership in Connecticut, New

York and nearby states, available records do not indicate the extent of slave ownership by the Holmes family. Twelve years later, however, Nicholas and Phoebe headed for Vermont, which became a hotbed of abolitionist sentiment.

Vermont Quakers were at the forefront of the anti-slavery movement in Vermont. They took a public stance in 1835 when they petitioned the state legislature to pass a resolution asking the U.S. Congress to abolish slavery in the District of Columbia. The above mentioned Higbee wrote, "This state was never a safe place for a slave owner to bring his 'property,'" noting that, "it was a Vermont judge who demanded a 'bill of sale' from the Almighty as evidence in his court."[10]

Later, Vermont Quakers, including the Robinsons of Ferrisburgh, hosted several stops on the Underground Railroad along the north-south corridor near where the Holmes family settled.

We know from a ground-breaking book by Elise Guyette, *Discovering Black Vermont*, that there was a meaningful presence of Black people in this part of Vermont in the 1800s. She wrote about Vermont's slaveholding past and about how freed slaves made a life for themselves.

A prominent example was the collection of Black families that resided on the Hill in Hinesburgh, 1790–1890. Hinesburgh (now called Hinesburg) abuts Charlotte to the east. Census data cited by Guyette indicates that the towns of Hinesburgh and Ferrisburgh included as many as 30 African-Americans in the first half of the nineteenth century.[11]

What about Charlotte? Did the Holmes farm employ Black people during its 100-year history?

According to Guyette, a few Black families resided in Charlotte in the 1860s.[12] One couple was married in the Charlotte Congregation Church in 1857, where the Holmeses worshipped.[13] There is some evidence that the Holmes farm employed Black people. Pictures of the apple-picking crews in the late 1900s and early 1900s seem to show 2–3 African-Americans, but the old pictures are hazy. At any rate, the family's slave-owning was in the distant past.

Evidence of the family's aligning with the rights of Black people are the act of manumission of young Nathaniel in 1776, the move to a section of Vermont dedicated to abolition, attending the abolition-influenced Quaker meeting house in Ferrisburgh, and attending a Charlotte church that held a wedding of a Black couple.

## B. The Johnses Settle Huntington

**Johns line of descendants from England to Huntington, VT**

1. Stephen (unknown life span) from England to Boston in 1703
2. Benjamin (born 1707) moved to Manchester, VT
3. Benjamin (born 1731) lived in Manchester, VT
4. Jehiel (born 1756) moved from Manchester VT to Huntington, VT, in 1786

The Johnses were Americans to the bone. They escaped the constraints of life in England and came to the New World. They speculated in land. They joined the colonial army and fought against the British. They ventured into the forbidding wilderness of Vermont and founded a town. The trajectory from England to Huntington was as follows:

(1) **Stephen** Johns arrived at the Massachusetts Bay Colony in Boston in 1703 with his wife, Jane.

(2) Stephen's son, **Benjamin**, who was born shortly after landing in Massachusetts, became a land speculator and later moved to the town of Manchester in southern Vermont. Benjamin's son, also named Benjamin, was born in 1731.

In 1775, just before the outbreak of the American Revolution, the elder Benjamin signed the "Association Test," which stated his commitment to "adopt and endeavor into execution whatever measures may be recommended by the Continental Congress, or resolved upon out Provincial Convention, for the purpose of preserving our constitution and of opposing the several arbitrary acts of the British Parliament."[14] Signing the document was seen as a test of loyalty to the colonial cause, but the family was already on board. They were dedicated patriots from the time of their arrival in North America.

(3) The younger **Benjamin** married Eunice Seeley and produced five children before he died in 1761 of smallpox. Benjamin's son, Jehiel, was born in Amenia, New York, in 1756, just five years before his father died.

(4) Thirty-two years later, **Jehiel** Johns founded Huntington in the Vermont wilderness. His story borders on the mythical.

After the death of his father, Jehiel received attention and care from his grandfather, the first Benjamin, and his uncles, Stephen and Daniel. An entry in The Colonial Genealogist by Dorothy Mae Johns describes Jehiel's early years:

> He moved about with them and probably benefitted from the experiences they all had together. And, above all, he inherited the pioneer spirit which had come down to him from his ancestors dating back to Edward Fuller, a passenger on the first voyage of the Mayflower.[15]

Just 20 years old, Jehiel signed up to fight in the Revolutionary War and was assigned to the regiment of Colonel Seth Warner. Jehiel appears on the Vermont Revolutionary Rolls. According to a historical footnote, Jehiel met the fabled Ethan Allen in the 1760s, probably in the Manchester area. Jehiel surfaced in an 1876 biography of Allen by Zadock Thompson, in which Thompson reported Jehiel's assertion in a letter to him that "Allen about that time was on very intimate terms with that noted infidel, Dr. Thomas Young." Jehiel's letter said that Allen "studied Deism with Doctor Thomas Young until he thought himself a conjurer."[16]

In January 1786, at the age of 30, Jehiel married Elizabeth Sexton. Elizabeth was the daughter of an early settler of Manchester. The previous fall, with land speculation in his blood, Jehiel had purchased lot 58 along the Huntington River. The lot was located at the foot of the Green Mountains, southeast of the town of Richmond and 25 miles from Lake Champlain.

Two months after their marriage, Jehiel and Elizabeth packed up their belongings and headed north into Vermont. They connected to iced-over Otter Creek near Rutland and made their way down the river on the ice or on adjacent roads to Lake Champlain west of the town of Vergennes. The map below (Figure 2.2) shows how Otter Creek cuts a swath through west-central Vermont.

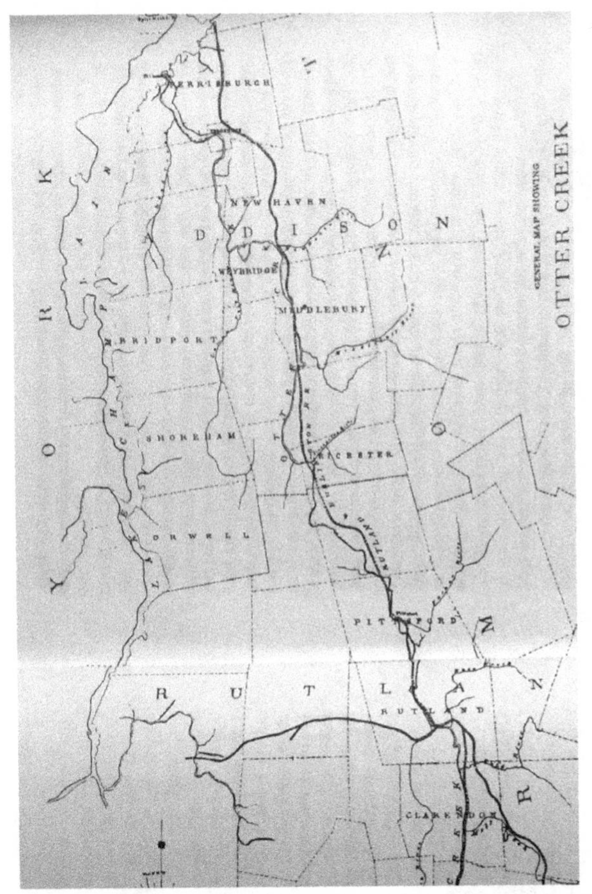


*Figure 2.2 Map of Otter Creek, Courtesy of the National Archives*

From there they traveled the ice north to Burlington and landed at the mouth of the Onion River, later named the Winooski River. At the time Burlington was settled by just three families. Jehiel and Elizabeth then traveled by land along the river to what is now Richmond.

On the map below (Figure 2.3), one can trace the route from the mouth of the Winooski River on Lake Champlain (north of Burlington) to Richmond on the Winooski, then up the Huntington River to the site of Johnses new homestead.

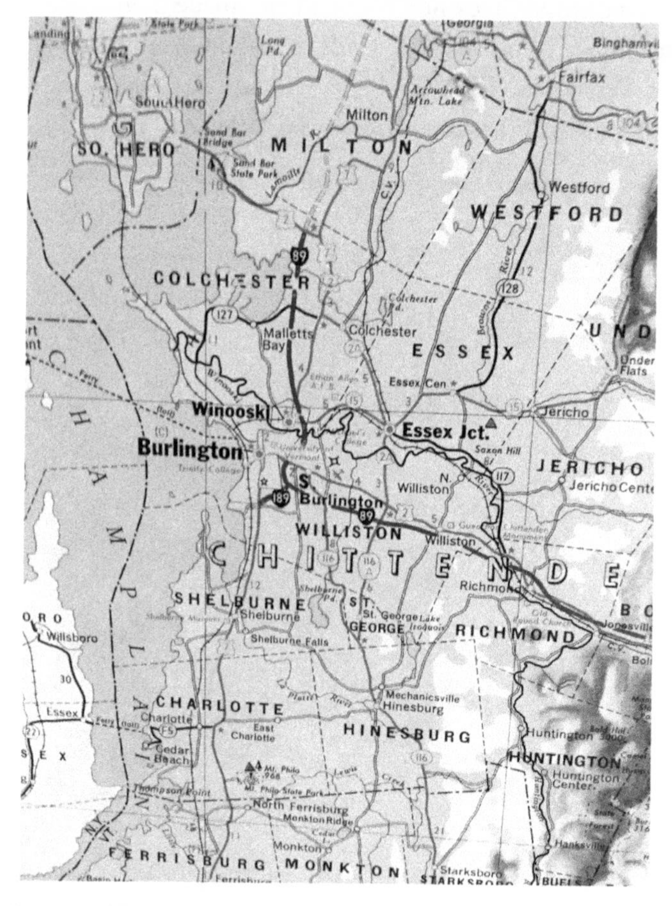


*Figure 2.3 Map of Winooski River, The National Survey, Chester, Vermont*

Jehiel deposited his wife and belongings in Richmond with Joel Bronson and his family and headed south into the wilderness to find Lot 58 and establish a 180-acre homestead.

The land was an unpopulated ocean of trees, reaching west and east of the narrow river. As dark and closed-in as the Huntington landscape may have appeared to Jehiel in the fall of 1786, a redeeming feature may have been the view to the east of Camels Hump (Figure 2.4—picture taken in the spring of 2021). The mountain can be seen from every direction in northern Vermont, but looms in a spectacular way over Huntington.

*Figure 2.4 Camel's Hump, © Emily Anderson*

His son, James, wrote this account of Jehiel's labors, which appeared in Abby Maria Hemingway's *Vermont Historical Gazetteer* (1867):

> (Jehiel) proceeded, with axe on shoulder and such other necessities as new settlers require, by marked trees through the woods to his pitch in the then un-broken wilderness of New Huntington and where he proceeded to fell trees over some two or three acres. And then lay up the body of the log-cabin . . . rolling together, notching and laying up the timbers of the lower half unassisted by other human mortal.[17]

The cabin was located on the Huntington River in what is known now as the Lower Village.

> Johns chose the site of his log cabin carefully, bearing in mind the needs of future family, and the

> development of the farm. Though the bed of the Huntington River is now a least twice as wide as it was at the time of the settlement, the stream was close enough to make the carrying of water for the household relatively easy. the high west bank of the river—long known as "Jehiel Johns' Slip Bank"—sheltered the spot from cold winds and drifting snow. Across the valley the land rose steeply to the east, thus providing additional protection.[18]

Once the cabin was completed with the help of two acquaintances from Richmond and made habitable, Jehiel returned to the Brownson abode to gather up Elizabeth and their belongings. On Nov. 13, 1786, Polly was born, the first of six children.

Jehiel and Elizabeth traveled back to the Brownson's by horseback for the birth. The Johnses lived in the cabin until 1806 when they built a frame house on the north end of the property.

Unmistakably, 1786 was an eventful year for Jehiel, who was just over 6 feet tall and described as "a bold, hardy, athletic man." He married, brought his wife north from Manchester by foot to Richmond, cleared land and built a log cabin in Huntington, established a household with Elizabeth in the new cabin, and saw the birth of his first child.

Jehiel went on to serve as moderator of the first town meeting of Huntington, the first justice of the peace, and in 1791 the first representative to the state's Legislature. He died in 1840. Jehiel's gravestone has the following epitaph at the bottom:

> *Life's longest term must soon depart*
> *The dying hour must come at last*
> *Tho' spared for four score years + four*
> *This grave will show I am no more.*

Jehiel's grave in the old Huntington cemetery is on the left in the row of family stones (Figure 2.5).

*Figure 2.5 Johns family gravestones, Huntington © Emily Anderson*

This eloquent statement on the tombstone was composed by his literary son, James. Elizabeth, who had traveled the same routes as Jehiel, lived until 1851. Her son wrote this epitaph for his mother, ending with a sobering last line:

> *First of my sex brought to this town,*
> *To keep a house was I*
> *Here by my partner I'm at rest*
> *For we were born to die.*

It is evident from the above picture in Huntington and the Quaker burial ground in Monkton that the cemeteries were set on beautiful sites away from the town centers. It may be purely coincidental that the burials in the 1800s of the early Holmes and Johns settlers (e.g., Nicholas and Jonathan Holmes; Jehiel and James Johns) overlapped with the "rural cemetery" movement" across New

England. The movement, inspired by the Transcendentalists, was launched with the creation of the Mount Auburn cemetery outside of Boston in 1831. As one writer pointed out, "The associations of the picturesque rural site would instill healing truths, of natural death and rebirth, in the cycle of the seasons."[19]

I remember my grandparents taking me to view the headstones of my Vermont ancestors. The visits were an occasion to tell stories—not sad stories—about my ancestors and their lives and to situate me in the family's history. Standing amidst the greenery of the cemeteries, the stories conveyed the cycle of life, death, and the birth of new generations. I was a new generation connected to the past.

James Johns, Jehiel's son, was a prolific writer and chronicler of life on the Vermont frontier. At age 13, he created a one-page newspaper, the *Huntington Gazette*, printed on one side of brown wrapping paper. In 1832, he founded The Vermont Autograph and Remarker, typically containing 1,500 words and six or seven articles authored by James. He wrote a history of Huntington, numerous poems, and continuous commentary on local affairs. His writings can be found in numerous libraries throughout New England. This picture of James (Figure 2.6) suggests a serious man with a lot on his mind.

*Figure 2.6 James Johns, Courtesy of the Vermont Historical Society*

As an appreciation of what his father had accomplished, James wrote a lengthy poem, "First Settlement of Huntington, with an account of a Bear and an Ox."[20] It ended with these lines:

*With gun and ball and powder arm'd*
*Himself and house to guard*
*Few of us yea indeed I doubt*
*If there be anyone*
*Of the young folks who realize*
*What their old sires have done.*

The May–June 1977 edition of *Vermont History News* reproduced a page from the Remarker, dated Feb. 12, 1849, and pointed out that "For over 40 years James Johns, an eccentric poet, author, and editor, produced an unusual pen-printed newspaper, whose small size, sharp prose, personal air, and often unconventional editorial positions made it unique in Vermont publishing history."[21]

Three generations after Jehiel Johns settled Huntington, his great grandson, Charles Johns, married Cornelia Ellison, of Charlotte. Their daughter, Rena, was born in 1882 and married Robert Holmes in 1911. Rena and Robert created a household at the Holmes farm on Lake Champlain, and soon thereafter built their own house on the property. The Holmes and Johns families stayed close for years to come. Rena's brother, Charles, shared a love of horses with Robert, and the two raced trotters together at state fairs and at famous racetracks, including Saratoga.

## C. The Huntington Rosses

Chester and Harry Ross were among the earliest settlers and farmers in Huntington. Catherine Ross, a descendant and also great granddaughter of Jehiel Johns, grew up in Huntington. Catherine married and became a career-long educator. Catherine taught in one-room schoolhouses in Huntington and nearby towns. With a deep commitment to academic pursuits, she made sure that her only son, Jacob Ross, received an excellent education.

Jacob attended high school at Montpelier Seminary, graduated from the University of Vermont in 1904, then received his MD at the University of Vermont Medical College in 1908. He became a general practitioner in Richmond before moving Middlebury to continue his practice, where he also became the first director of physical education at Middlebury College. He served with distinction in the Medical Corps during World War I in France as a flight surgeon in 17th Aero Squadron. While in France and later he contributed regularly to the medical literature.

Upon Catherine's death, Jacob inherited a section of land from his mother in the north end of Huntington, which belonged to her family for several generations. He decided to honor Catherine by leasing to the town an acre section of the property for the construction of a school at an annual rent of one cent. He envisioned a "modern" two-room school that would serve the children of Huntington.

The Catherine E. Ross School (Figure 2.7) opened in 1915 with running water, a furnace, and sizeable meeting spaces. The school burned to the ground a year later but was rebuilt the next year on its old foundation and reopened. The school served the youth of Huntington until 1965 when a new, larger school was opened at another site. In 1966, Jacob's four children—Katherine, Austin, Charlie and Helen—deeded the land to the town.

*Figure 2.7 Catherine E. Ross School, From Huntington, 1976, unattributed*

## D. Ties That Bind

The Holmes, Johns and Ross families came together in Vermont eight generations after Francis Holmes arrived in the New World in 1643. Between 1881 and 1885, C.T. and Clara Holmes had four children, including Hannah and Robert. Hannah married Jacob Ross. Robert married Rena Johns.

Although the two marriages formalized the connection of the three families, it is likely that the three families knew each other over the previous years. Several Vermont towns are clustered to the west of Green Mountains (Richmond, Huntington, Hinesburg, Bristol, Monkton, Ferrisburgh, Charlotte), and members of the three families lived at different times in these different towns. Whether through church, schools, farming business or common avocations, people in these small towns often knew each other. When Rena married Robert, for example, Robert and Rena's brother, Charlie, had raced trotters together for many years at Vermont fairs.

Also, as often happens in rural communities, families relate in more than one way. The Jehiel Johns connection is an example. Catherine Ross was the great-granddaughter of Jehiel, and her son, Jacob, married Hannah Holmes. Rena Johns was Jehiel's great-great-granddaughter, and she married, Hannah's brother, Robert Holmes.

After a stint in Richmond, Jacob and Hannah moved to Middlebury, just 25 miles from the Holmes farm to the north, where Jacob had his medical practice. There was a strong interdependence among these branches of the family, especially during World War I when Jacob was stationed in Europe for two years and Hannah was at home with their young children. Chapter VI includes detail on the years of Jacob's service in World War I and how Hannah stayed close to her relatives back at the farm. When the farm failed in 1923, the Holmeses moved to Middlebury and the bond between the two branches became even closer.

## E. The Lake

There is another actor in this story: Lake Champlain. Its weather, its 120-mile expanse, its beauty, its benefits, its dangers. It is a fact of life for those who live along its shores.

As it has for families since the 1770s, the lake shaped the life of the Holmes family in a fundamental way. I moved to the shores of Lake Champlain in April 2018 and have experienced the same vistas and forces of nature as my ancestors, including the iced-over surface and the stiff winds sweeping down from the Adirondack range.

As I am sure it did for the Holmes family when they made the trek from Monkton in 1822, it took a few months for me to adjust to the ever-changing weather along the lake. Figure 2.8 shows the west view from the site of the Charlotte farmhouse built in 1912 by Robert Holmes. The picture was taken in November 2020, but the vista has not changed over all these years.

*Figure 2.8 View of Lake Champlain*

When the Holmeses came to Vermont in 1788 and 34 years later went on to the lake, memories of the Revolutionary War were still vivid among older Vermonters, including members of the Holmes family. Several had fought in the colonial army. The move to the lake in 1822 brought the family to a body of water where decisive battles were fought, especially during the autumn of 1776.

Lake Champlain is a narrow corridor, 120 miles long and 12 miles at its widest, and one can imagine the scene in 1776 when Benedict Arnold's over-matched navy fought a defensive battle against the British. The fight extended from the Canadian border to Crown Point, some 90 miles to the south.

Arnold, who is remembered today as a traitor, was assigned to build the colonial navy in a shipyard located at what is now known as Whitehall, New York, just off the south end of the lake. Although the British defeated Arnold's small navy as winter approached in 1776 (Arnold burned what was left of his navy in a small cove in Panton), he delayed the British advance long enough to foil their plan to divide the colonies on a line from Canada, down Lake Champlain to the Hudson River and on to New York City. It is widely accepted that Arnold—and oncoming winter— saved the American cause. By the following spring, the colonists had strengthened their forces, and the war raged on.

Today, one can look out on the lake and picture the movement of canoes, small boats, and naval vessels up and down the lake. During wartime—the French and Indian War, the Revolutionary War, the War of 1812—there were armadas of French, British, and American ships carrying heavy armaments and thousands of troops with colorful flags flying. Even now, standing on the shore at the site of the Holmes farm or where Arnold sank and burned his remaining vessels, one feels close to history.

As Ralph Nading Hill pointed out, Lake Champlain has remained remarkably unchanged over the last 200 years.[22] The lake is largely untraveled, except for a few locations near Burlington, the marinas that dot the lake, and the ferries. The view to the west—the majestic Adirondacks—is still thrilling.

Most importantly, the lake is always *there*, not ever to be ignored; the seasons have the same rhythms; and one experiences the same mix of weather conditions, from sublime summer days to winter days of frigid gale force winds and ice. Beginning as a doctoral student, Vermont historian Nicholas Muller wrote extensively about the history of the lake:

> Over the years you begin to feel the lake; its moods and colors, its seasons, its tastes. It affects you as an individual. You begin to get a real fondness for it; you get a sense of awe when it frightens you—and it does from time to time—and a real sense of well-being and peacefulness from it.[23]

Like me, Muller was drawn to the lake and moved to its shores later in life. Moreover, from what I saw in my grandparents, my father, and his sister, each of whom lived at the farm until 1923, the lake had a lasting emotional impact. They loved it, returned frequently to its shores, and had countless lake stories to tell. There are many newspaper clippings saved in the Holmes family archive for more than a hundred years, including the following undated poem:

> *The hazy hills are dreaming*
> *Majestic dreams of days of old;*
> *Their hoary brows are gleaming*
> *Inspired with the sunset's gold.*
> *Then come, ye worn with cities strife,*
> *And health of heart and soul re-gain*
> *When heaven and earth with peace are rife,*
> *In Burlington and Lake Champlain.*[24]

I realize now that it was a bittersweet moment in the 1950s when they took me, a kid, on picnics at Button Bay State Park near Vergennes, a few miles south of the old farm. Knowing the unchanging nature of the lake, to visit in the 1950s or live along the lake in 2021, is to know something of what it was like for Nicholas, Jonathan and C.T. to live there in the 1800s.

# Chapter Three

# Entrepreneurs of the Land

*The 101-year history of the Holmes farm is a striking example of entrepreneurship in rural Vermont.*

## A. The Property

The move of Nicholas Holmes to the shores of Lake Champlain was a land exchange. Nicholas sold his Monkton property to Joseph Hurlbut and, in turn, acquired 215 acres of land on the shores of the lake from Hurlbut on April 17, 1822.

The earliest land records of Charlotte are located in the office of the town clerk, and one can inspect the 1822 deed, written in longhand, for Nicholas's purchase. The deed for the Charlotte land stipulated a sale price of $6,000. The opening and closing sentences of the ancient deed, crinkled and difficult to decipher, are shown in Figure 3.1. The deed was witnessed by Nicholas's brother, Jonathan, and Stephen Haight, a relative, and officially recorded by the town clerk the following day.

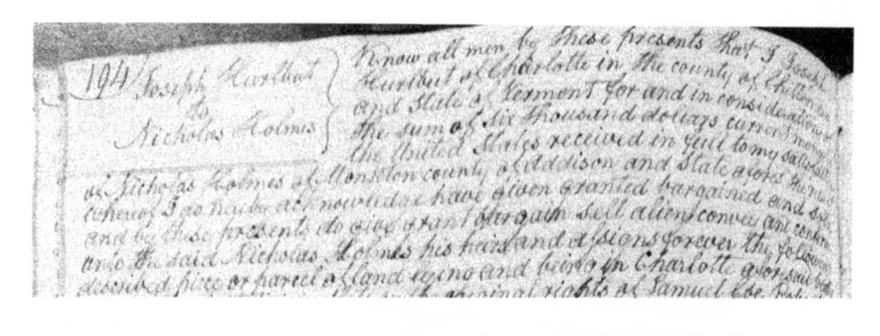
194 Joseph Hurlbut to Nicholas Holmes

Know all men by these presents that I Joseph Hurlbut of Charlotte in the county of Chitten[den] and State of Vermont for and in consideration of the sum of Six thousand dollars current money of the United States received in full to my satisfaction of Nicholas Holmes of Monkton county of Addison and State aforesaid the receipt whereof I do hereby acknowledge have given granted bargained and sold and by these presents do give grant bargain sell alien convey and confirm unto the said Nicholas Holmes his heirs and assigns forever the following described piece or parcel of land lying and being in Charlotte aforesaid

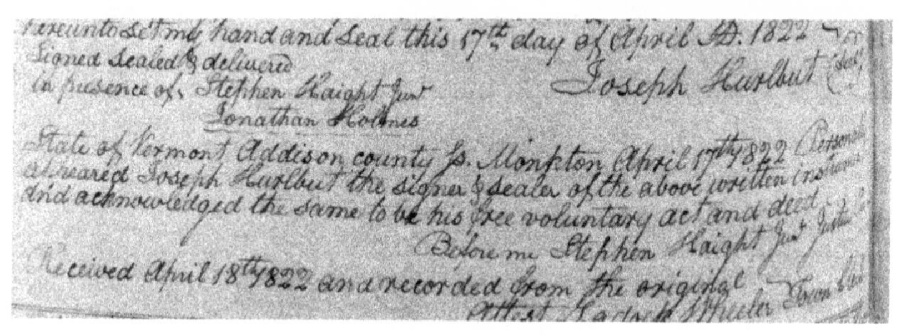
hereunto set my hand and seal this 17th day of April AD. 1822

Signed sealed & delivered in presence of Stephen Haight Jun. Jonathan Holmes

Joseph Hurlbut (seal)

State of Vermont Addison county ss. Monkton April 17th 1822 Personally appeared Joseph Hurlbut the signer & sealer of the above written instrument and acknowledged the same to be his free voluntary act and deed

Before me Stephen Haight Jun. Justice

Received April 18th 1822 and recorded from the original

*Figure 3.1 1822 deed*

The 32 other deeds for land transactions through the family's departure in 1923 can also be found in the Charlotte town records. Looking at the fraying Charlotte Grand List for 1822 stored on a high shelf in the town clerk's reading room, it is reported that the Holmes farm had 115 acres. This many years later, it is impossible to explain the discrepancy between the acreage in the 1822 deed of sale—215 acres—and the Grand List total of the same year.

Both before and after the sale of the land to Nicholas, the Hurlbut family were active landowners in Charlotte with numerous land transactions. It appears that Joseph had acquired the land purchased by Nicholas from a Hurlbut relative around 1800. In Monkton, the Hurlbut family occupied Nicholas' old property for several generations but suffered a terrible tragedy. William Higbee, writing in the 1889, recalled the incident:

> Mr. Hurlbut and his wife Thalia Dean started together for the west in 1856. Mr. Hurlbut taking a fine pair of Morgan horses that he intended selling

> before his return. While crossing Lake Michigan, the steamboat burned and both Mr. Hurlbut and wife drowned. The bodies of Mr. and Mrs. Hurlbut were never found. The sudden and shocking calamity cast a great gloom over the entire community.[1]

Among the mourners, we presume, were members of the Holmes family.

The new Holmes property was on the northwest shore of what became known as Holmes Bay, reaching around the westerly point, then farther south along Lake Champlain. The move from the foothills of the Green Mountains in Monkton was a move to the Champlain Valley. This part of Vermont, stretching south from the Canadian border, was perceived to be promising land for farming. In a book exploring the archaeological and anthropological origins of Vermont, William Haviland and Marjory Power offered this description:

> The soils of the Champlain Valley consist of water-deposited sands, silts, and clays, with glacial drift at higher elevations. These are some of the richest farmlands in the state and are easy to cultivate with simple hand tools. These soils, a growing season of 130 or more frost-free days, and an annual precipitation of 33 inches made the valley suitable for native horticulture.[2]

The Vermont Department of Agriculture, writing in 1897—75 years after the arrival of the Holmes—described the Valley as follows:

> The surface of this section is quite level, and the soil is very strong and fertile. The soil is largely clay, and a mixture of clay and loam. It noted for abundant crops of hay and grain. The system of farming here, as well as the surface of the country, somewhat resembles the prairie states of the west. Nearly the whole of this section is especially adapted to the

> growing of fruit. The apples grown here are known as the finest known in the markets, and are grown with but little expense, as compared with many localities. The soil is naturally so well supplied with the elements of fertility, that but little artificial fertilizing is necessary.[3]

The rosy picture in this and other promotional documents disseminated by state government did not tell the full story, of course. Creating and sustaining a profitable farm is always a challenge. Yet, from the early 1800s until today, the Champlain Valley has been the locale of hundreds of flourishing farms and orchards. Nicholas' instinct to move his farming to the shoreline was a good one.

The property of James Hill, just to the north of the Holmes property, was one of the first homesteads in the Town of Charlotte, settled in 1784. The Hill homestead stood at Hill's Point, with Hills Bay to the north. Figure 3.2 shows the location of the Hill and Holmes properties.[4]

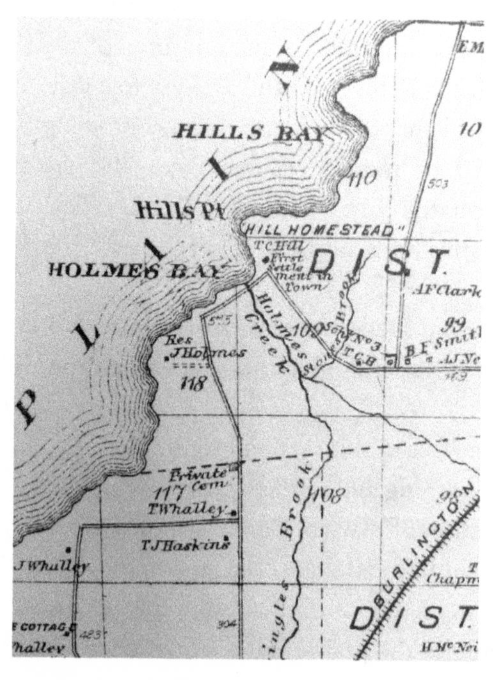


*Figure 3.2 Map of Charlotte shoreline, F.W. Beers Atlas of Chittenden County, 1869*

Hill was married to the daughter of Vermont Gov. Thomas Chittenden, who played a prominent role in Vermont history. The History of Chittenden County, published in 1886, reported the following about the Hill's:

> Hill's wife is said to often declared that she was for three months, "The handsomest woman in town, for the very reason that she was the only one." Being a man of some means, James Hill erected a grist-mill by the creek near the house, which long ago disappeared, but which for years supplied the demand of the farmers in the neighborhood, even those who lived on the other side of the lake. He afterwards bought the grist-mill in Ferrisburgh.[5]

Today, if one looks south from Holmes Creek, one sees the north-facing shore of the former Holmes farm. Figure 3.3 shows the shore line on a winter day, iced over.

*Figure 3.3 North shore of Holmes property*

## Buildings

The Holmes house, which was on high ground, faced south and became Nicholas's homestead with his wife, Sarah, and his children. The house (Figure 3.4) was constructed with bricks made in the Holmes brickyard on the property.

*Figure 3.4 First Holmes house*

The front porch (Figure 3.5) was a place of relaxation for the family.

*Figure 3.5 Porch of house*

When Nicholas's son, Jonathan, was married to Hannah Smith, the new extended family of Jonathan and Hannah and their children, joined the clan in the house. Later, Jonathan's son, C.T., and his family, also moved into the house, making it three families under one roof.

In the 1880s, a large frame house was built to the south along the lane that split the property on a north-south line. It housed Jonathan's son, William, and his sizeable family of eight. William's wife, Mary Will, described her new home in a letter to her sister:

> We are living in the new house on the hill south of Father Holmes (Nicholas' son, Jonathan). Perhaps you remember the hickory trees in the meadow south of the barn. They are directly in front of our house. The house is far from completed, but it is home. There is a sitting room facing south with a bay window, two bedrooms on the north side, a kitchen and a small pantry plastered, and three small chambers, one over the kitchen.[6]

A third house (Figure 3.6) was built in 1911 at the time Nicholas's great grandson, Robert, married Rena Johns.

*Figure 3.6 House under construction*

The house overlooked the lake to the west. The photograph shows the house under construction. A few years later, we see Rena and her young children, Marion and John, next to the finished house, with cold frames behind them for seedlings to be transplanted in late spring (Figure 3.7).

*Figure 3.7 Rena and children*

The Thurber family, who took over the farm property in 1936, rebuilt the house on the original foundation in 1949. The next owners, in conjunction with the Charlotte Fire Department, burned down the Thurber house and built a substantial lakeside property at the same site.

A large pier was constructed on the north shore to provide for the loading and shipping of apples, which enabled the Holmes to ship apples to U.S. cities and to foreign ports like London. Apples were also shipped by rail, and an apple storage house stood by the Charlotte tracks. When they shipped apples by rail during the winter, a farm employee rode in the box car and stoked a coal stove to keep the apples from freezing.

The photograph below (Figure 3.8) shows the main buildings of the farm, circa 1880.

*Figure 3.8 View of property*

From the beginning, the family constructed buildings to support the various activities of the farm. The farm property was not fully developed until the 1880s, however, when the apple-growing and the horse-breeding operations came to maturity.

Marion Holmes, who was born on the farm in 1912 and lived there for the first eleven years of her life, sketched the property in 1980 as she remembered it. She identified:

- Substantial acreage—two parcels—for the orchard
- Half-mile racetrack for the trotters
- Two primary pastures
- Three houses
- Two pump houses for irrigation (water pumped from Lake Champlain)
- Horse barn, cow barn, sheep barn, chicken house
- Apple shed for storing apples in barrels, with a machine shop at one end
- Corn crib

- Ice house
- Blacksmith shop
- 2–3 boarding or "tenement" houses for hired hands (a third house may have existed)
- Burial ground

Below (Figure 3.9) is a map of the property based on Marion's sketch.

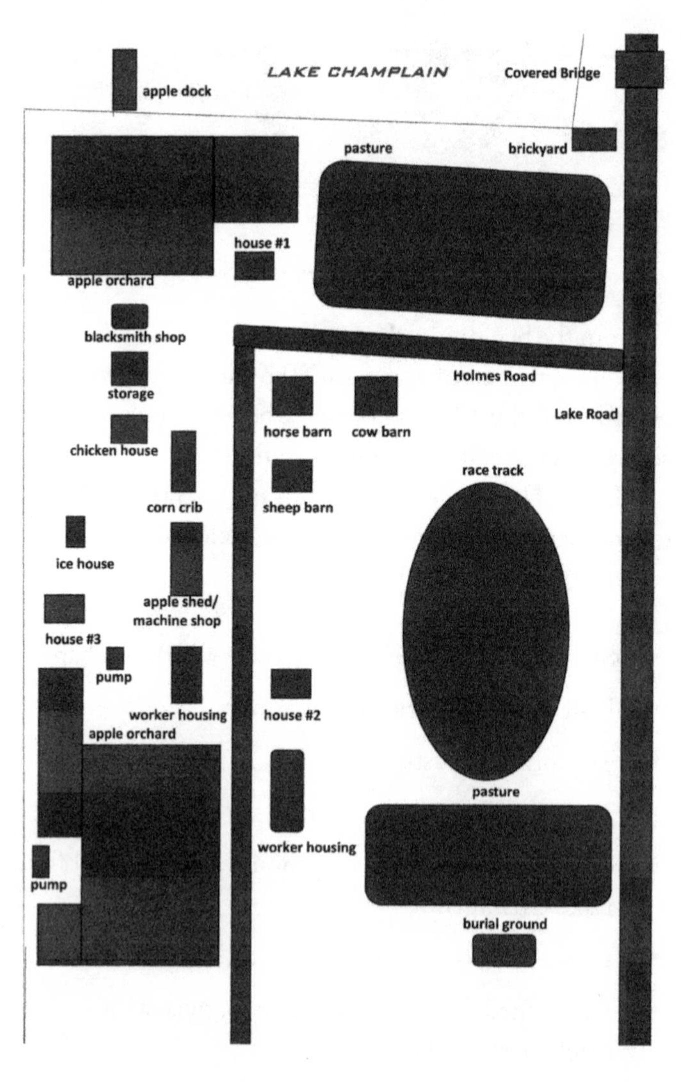


*Figure 3.9 Map of property*

The property was bordered on the east by Lake Road, running north through the covered bridge. The first apple shed of 100 feet in length was blown down in a fierce wind and rebuilt.

### *Mystery of the Burial Ground*

The site map includes the location of what Marion Holmes's drawing labeled a "burial ground." There is no evidence suggesting that this was an Indian burial site. Because the headstones for the family who lived on the farm are located in the cemetery behind the Charlotte Congregational Church and not here, I initially surmised that this was the burial location for animals who died of old age or were put down. Over the 101-year life of the farm, the family owned hundreds of animals—horses, cows, sheep, oxen. Assuming that dead animals were not burned, pushed into the lake or taken away, the farm needed a burial site. There is another explanation, however.

Based on word-of-mouth passed down over the years among nearby residents and a close inspection of the Charlotte-area map in the 1869 Beers Atlas, it is evident that this site on the southeast corner of the property was not for burying animals. The map shows clearly a "Private Cem" at this site just north of the Whalley homestead. In fact, family members were buried on this small hill. Walking the ground, one sees indentations where headstones were previously located and a single massive headstone, fallen flat on the ground.

This headstone has the names of Nicholas and Sarah, the founders of the farm who made the move from Monkton to Charlotte in 1822. Nicholas died in 1863, and his wife, Sarah, died in 1866. Cleared of mud and debris, the fallen headstone is pictured next in Figure 3.10. As for dead animals, the family probably dug pits at flatter locations on the property and buried animals there. In time, the animals decomposed.

*Figure 3.10 Headstone of Nicholas and Clara*

There may have been as many as ten relatives buried on the ground. Lacking old pictures, one must imagine the solemn ceremonies, presided over by a local minister, that took place on this hill at the far end of the property. Yet, even this part of life at the farm did not last.

When the farm failed and the family moved away, the caskets, headstones and bones of the family were dug up and transported a few miles away to a new Holmes plot behind the Charlotte Congregational Church. Probably because of its size and weight, the large headstone for Nicholas and Sarah was left behind. Perhaps for the same reason, Jonathan's headstone never made it to the Congregational Church. Jonathan's headstone is embedded in the floor—upside down – in front of the fireplace of the Allen family house, which is just below the cemetery on Lake Road.

In addition, a story handed down over the years suggests that, when they did the digging in 1923, they could not find the body of a baby or small child. This may have been Mildred, daughter of C.T. and Clara, who was born in 1889 and died three years later. As for Nicholas, Sarah, and Jonathan, their caskets were taken to the new resting ground where another headstone memorialized their lives.

Knowing that others would take over this private property, the family didn't want to leave its dead behind. Thus, the day came in 1923 when the caskets were dug up and transported across town. This had to be a deeply sad moment.

### *Other Aspects*

In 1859, with ownership of the farm now shown as "Nicholas Holmes & Son, " Nicholas' son, Jonathan, added about 32 acres of land to the farm by purchasing Birch Island in Converse Bay near Thompson's Point from William Yale. The island was later renamed Putnam's Island and then Garden Island. With a flock of 100 sheep at its largest, the Holmeses pastured their sheep on the island in the summer by wading or swimming the sheep from the shoreline to the island.

With the apple orchard emerging as the main business of the farm, Jonathan sold the island to Henry Putnam, Jr. in 1887 for $2,000. The Burlington Free Press reported on Oct. 28 that Putnam, a wealthy resident of New York City "proposes building a fine cottage on his island and will take great pains to beautify the grounds."[7]

In 1899 Putnam hired Albert Williams, a local resident, to be caretaker, and Albert and his wife, Sarah, moved full-time to the island and lived in a farmhouse, now called Birch Cottage. They watched over the several cottages, the animals, a garden and fruit trees. Their children attended school on the mainland, and they often boarded with local residents during the winter when traveling over the ice by horse-drawn sled was a daunting challenge.[8]

The Free Press article went on to point out that real estate experts believe that "Lake Champlain is to become one of the most popular summer resorts in the United States . . . It is understood that several parties have been negotiating for lake front property at high figures." This was an accurate prognosis. In fact, in the years after the Holmes farm was evacuated in 1923, the shoreline property became a pricey location for private homes and summer residences.

Adjacent to the property on the northeast side and crossing what became known as Holmes Creek (flowing from the east into the lake), a bridge was constructed about 1840. The bridge was replaced around 1870 with a covered bridge called the Holmes Creek Bridge. Figure 3.12 shows the bridge looking north, with the Hill homestead in view.

*Figure 3.11 Covered bridge, looking north*

The next photograph (Figure 3.12), circa 1915, looks south through the bridge and shows the first Holmes residence in the background.

*Figure 3.12 Covered bridge, looking south*

The bridge, which still stands, is 41 feet long and 12 feet wide. The width was specified by the town selectmen to provide passage of a "load of hay, high and wide."

The bridge, reportedly the shortest covered bridge in New England, is known for its rare tied-arch truss, which is constructed of five thicknesses of planks bolted to the framework of the bridge. There is a tie—or stringer—below the deck that runs the length of the bridge with the ends supported by abutments. The interior construction and the beautiful arched beam are shown in Figure 3.13.

*Figure 3.13 Covered bridge tie © Emily Anderson*

Figure 3.14 shows the plaque that is placed inside the bridge and references the Holmes apple orchard.

*Figure 3.14 Covered bridge plaque © Emily Anderson*

The incessant pounding of the waves off Lake Champlain has had its effects over the years. In a document located in Charlotte town records, dated Aug. 17, 1863, town moderator Peter Van Vliet Higbee warned a town meeting to deal with the advancing waters that were overtaking the lakeside road to the north of the Holmes property and destabilizing the original bridge across Holmes Creek. The meeting agenda included these action items:

1. To see what action the town will take in regards to fixing the road and bridge.
2. To see whether the town will open a new road from John Holmes to the place formerly owned by James Hill.[9]

Town citizens decided to move both the road and the bridge 40–50 feet to the east to escape the advancing waters. A few years later, in about 1870, the bridge was replaced with the covered bridge named for the Holmes family. Even then, however, the ravages of the pounding surf and strong winds had a deleterious effect. Mary Holmes, born on the farm in 1861, remembers as a child her parents getting her up to see the covered bridge go out with the wind and ice. [10]

The new bridge was rebuilt, and traffic along Lake Road was resumed.

There was once a road on the shoreline extending southwest from the covered bridge along the north shore of the Holmes property. This road was overtaken by the receding shoreline. Likewise, Birch Island, once estimated to be 40 acres and owned by the Holmes, is now significantly reduced in size.

Gone is an old oak tree on the north shore near where, according to tradition, there was an annual Indian encampment. We know that there were Indigenous peoples' settlements along Lake Champlain and inland prior to the 1770s, and those who came north found evidence of earlier occupation. In the 1760s when English settlers began to pour into Vermont, hundreds of acres in the Champlain and Connecticut

Valleys were found already cleared of trees, where people had grown their crops.[11]

In their book, *The Original Vermonters: Native Inhabitants Past and Present* 1981), William Haviland and Marjory Power substantiate this early history and the presence of the Abenaki to the north of Charlotte near the Winooski River and to the south of Charlotte at the mouth of Otter Creek.[12] Although the settlers perceived places like Huntington as *terra incognita* (heavily forested with no evident settlements), there is substantial evidence of Indian presence along the Champlain corridor and near the rivers flowing into the lake.

Charlotte residents have found hundreds of Indigenous artifacts over the years, including arrow heads and tools. A sampling of these artifacts, see Figure 3.15, are displayed at the Charlotte Historical Society Museum.

*Figure 3.15 Abenaki artifacts*

Although there is no evidence to suggest that the Abenaki cleared land at the site of the future Holmes property, it is likely the Abenaki hunted and fished near the mouth of Holmes Creek.

As the Holmes generations evolved, farm ownership evolved, with various acreage entries in the Charlotte Grand List:

***1822:*** Nicholas Holmes – 115 acres

***1843:*** Nicholas Holmes & son – 205 acres

***1863:*** John Holmes – 205 acres

***1880:*** Jonathan Holmes and sons – 200 acres

***1882:*** C.T. and William Holmes – 114 acres

***1883–1892:*** (same listing each year):

C.T. and William Holmes – 114 acres

John and sons –177 acres

William Holmes –33 acres

Island – 40 acres

***1894–1896:*** (same listing each year)

William Holmes – 105 acres

C.T. Holmes – 217 acres

***1900:*** C.T. Holmes – (no acreage listed)

It is difficult to connect the ebb and flow of acreage between the 1822 deed of sale (215 acres) and the Grand List totals over the years. The various land transactions recorded in the town clerk's office—33 in total—do not fully explain the Grand List totals. Perhaps of greater consequence is the Grand List of 1900 and subsequent years indicating C.T. as the sole owner. One hint of what happened came from C.T. in 1914 when he commented:

> By that time (c. 1900) there was a lawsuit came up in regards to the farm and it was a question whether to get off the farm or do something and redeem it. We went into the suit and I won out and got the place.[13]

It appears there was litigation between the Holmes family and nearby neighbors or, perhaps, creditors.

In 1898 William and his family moved to Proctor, Vermont, where he took a job with the Proctor Marble Company. In the face of financial challenges, did William conclude that the farm was not profitable enough to support both C.T.'s family of four and his family of nine, and he would need to make a move? Or, as a result of the litigation over possession of the farm and C.T.'s leadership role in handling the Holmes interest, did this leave William on the outside? Or, in the end, was it a matter of C.T. being more passionate than William about the farm and the orchard, which led William to pursue another avenue of employment and income?

The available records do not answer these questions but, most likely, it was some combination of these factors.

Regardless of what motivated the break, we know that in the ensuing years C.T. was the visible and forceful leader of the farm and aggressively built the orchard. We know also that the Charlotte and Proctor branches of the family remained close, with visits in both directions and large gatherings on special occasions.

## B. The Farming Enterprise

Virtually all of us are descended from farmers. Into the mid-20th century, most American families, even those who had moved to cities and towns, had experience and talent with horses, milk cows, backyard crops, using a scythe, canning, acquiring wood for heating, and other activities associated with a farm.

Then, of course, there were the necessary skill sets and areas of knowledge of those families who stayed on the farm. Fortunately, when Nicholas came from Monkton to the farm on Lake Champlain in 1822, he brought experience and knowledge for building and sustaining a farm.

Equally important, he brought an attitude and a way of thinking. Self-sufficiency was an obligatory attribute. An historian of Vermont farms captured this aspect:

> In the first half century after the Green Mountain
> Boys had destroyed British power in their particular

> section of North America, Vermont was a frontier state with the usual frontier characteristics of self-sufficiency and diversity in agriculture.[14]

This first half century of the 1800s was the time of Nicholas, and self-sufficiency personified Nicholas and later generations.

Equally important, entrepreneurial thinking was necessary for a farmer: a readiness to try new things, take a calculated risk, be analytical about how things are working, create one's own solutions, and envision the future. As this story conveys, Nicholas and succeeding generations, personified in particular by his grandson C.T., brought an entrepreneurial instinct and an experimental mindset to their work. This way of thinking coincided with an "era of improvement" in the late 1800s among farm families. David Donath in an article, "Agriculture and the Good Society," observed:

> The contemporary watchwords "progress" and "improvement" implied adopting scientific techniques and new technologies, taking advice from agricultural journals and the newly formed experiment stations, participating in agricultural societies and perfecting farm practices to improve both the quality and quantity of produce.[15]

As shown in Chapter Five, the apple orchard under C.T. was an example of these developments. Indeed, the 101-year history of the Holmes farm is a striking example of entrepreneurship in rural Vermont. We know too that the Johns branch of the family brought a similar mindset. George Read, a close relative of Rena Johns who married Robert Holmes, was well-known for breeding "new oat" and "new barley" at Read's Experimental Farms in Charlotte.

Known as a "hybridizer," Read elicited the praise of Harvard's Professor C.J. Pringle, a pioneering botanist: "One of the most capable and scientific hybridizers to carry on after me the work of plant breeding is Mr. G.A. Read of Charlotte."[16]

Frank Bryan, keen observer of Vermont of politics and culture, pointed out another dimension of farming that applied to the Holmes farm: the growing role of technology. In his book, *Yankee Politics in Rural Vermont* (1974), he points out that "Rural areas in the United States have always been machine filled." He also notes the misperception that "science and technological innovation have traditionally spawned in cities."[17]

> One need to go no further than Vermont to document the inventiveness of rural society. Before 1860, Vermont inventors produced the first steel square, rotary pump, platform scale, and American globes. A dozen years before Fulton, Samuel Morey of Fairlee, sailed a paddlewheel steamboat up the Connecticut River.[18]

Bryan's observations reflect the Holmes's century-long acceptance of the scientific ethic and seeking technological solutions to solve problems and attain greater efficiency. The family tinkered with how best to draw water from the lake, make bricks, utilize pumping technology to irrigate the orchard, apply manure to crops and the orchard, and find chemicals that work best to eliminate pests. C.T., as shown later, invented and patented a pruning saw for trees and utilized what he learned about mechanical pumps to become a sales representative for the Field Force Pump Company of Elmira, New York.

Another indication of scientific advancement on the farm came in the early 1900s. Agricultural experts, primarily in schools of agriculture and state departments of agriculture, began to study systematically the farm economy. A 1914 study, entitled, *What the Farm Contributes Directly to the Farmer's Living*, observed: "The three important elements furnished by the farm for the family are food, fuel, and the use of the dwelling."[19] Vermont's Lamoille County was one of ten sites featured in the report.

A 1925 study by the United States Department of Agriculture, entitled *The Family Living from the Farm*, looked at the budgets of several thousand farm families from 21 states, including Vermont. The data show that the farm contribution was approximately one-third of the cost of living of farm families and two-thirds of the cost of food, fuel and house rent.[20]

> The family living from the farm lends safety and stability to the farm business and to the farm life. It enables the farmer to reduce materially the cash cost of living and to tide over lean years and hard times that would be ruinous if he had to buy all the living for himself and family in the market.

This point—the important contribution of the farm itself to the budget of the farm—may be self-evident, but researchers began to put a dollar figure on this piece of the overall financial picture.

Another facet of the farm budget was the contribution of the women, particularly farmer's wives, to the income picture of the farm. A 1933 study by the Vermont Agricultural Experiment Station, entitled *Cash Contribution to the Family Income Made by Farm Homemakers*, discussed the imperative to find additional sources of income during the ravages of the Depression. Recognizing the traditional assumptions about gender roles (the men in the field; the women at work in and around the house), the study noted:

> Due to the decreases in farm income during the present period of depression there is a growing desire, and often a real need, for farm homemakers to contribute to the cash incomes of their families. Many women on Vermont farms, as elsewhere, find they must earn in order to maintain their accustomed plane of living, pay the taxes on the farm, repair the house, and educate their children.[22]

The study distinguished between the difficult-to-measure contributions to farm income (housekeeping, preparation of food, making and repairing clothing, childcare, etc.) and the homemaker's earning of "actual cash." It identifies poultry and home vegetable gardens as common sources of income. The Holmes women were a prime example of contributions in both realms: management of the daily life of the household and going forth to earn income. On the latter, as noted below, Hannah Holmes brought a keen business sense to her chicken business, sold eggs for cash in Burlington, and brought in modest but important income for the family.

In recent years, there has been a growing interest in acknowledging what a leading feminist scholar, Silvia Federici, calls "reproductive labor." The term refers to:

> Having children and raising them; it indicates all the work we do that is sustaining—keeping others around us well, fed, safe, clean, cared for, thriving. it's weeding your garden or making breakfast or helping your grandmother bathe—work you have to do over and over again, work that seems to erase itself. it is essential work that our economy tends not to acknowledge or compensate.[23]

This description captures much of what the Holmes women did for the family. But was it acknowledged as important? Or was it seen to be a secondary, mostly unskilled contribution?

On farms of this era, the male spouse did not head out each day to a job in town or a nearby city. Rather, the men and women on the farm were in close proximity on the farm's 200 acres, and there was day-long communication and contact. The family usually ate three meals together each day and shared the ups and downs of how each day unfolded.

For sure, men and women played different roles, but it appears that there was a foundation of mutual respect and a shared commitment to the fate of the enterprise. Nicholas, Jonathan and C.T. knew that the intelligence and organizational skills of the women were integral to the success of the farm, whether in managing the household, keeping the books, or orchestrating the affairs of the family.

According to what I heard and sensed from my relatives who lived on the farm, there was an acknowledged partnership between the men and women. In time, of course, many female offspring aspired to lives and vocations away from farming. Teaching became an accepted avenue for leaving the farm, at least until it was time to marry.

Shaped by an entrepreneurial instinct, an inventive mindset and a spirit of self-sufficiency, the Holmes farm became a conglomerate of multiple enterprises. The various enterprises served two purposes: To provide for basic needs of the family (e.g., the family consumed farm products like vegetables, meat, apples, and milk products) and to earn income from the sale of produce, animals and products.

In telling the story of the Holmes farm, a relevant question is the extent to which the farm typified Vermont farming during the 1800s and early 1900s. Although the thriving apple orchard and the horse business may have been unique features, the historical record indicates many elements were common with other farms. A 2019 book by Michael Foley looks back nostalgically to the "settled agriculture" of the nineteenth and early twentieth century. His book, entitled *Farming for the Long Haul: Resilience and the Lost Art of Agricultural Inventiveness,* describes characteristics that pervaded the Holmes farm:

> Farming is not first of all a business . . . It is a way of living. Either it satisfies a passion in us, a deep need, or we pretty quickly give it up. Those facts guarantee a sort of personal resilience.
>
> Any open discussion among farmers reveals some near constants: a passion for nurturing animal and plant life; a love of working outdoors; an appreciation for the mysteries of nature; a commitment to feeding people; a talent for troubleshooting; a fierce independence. We are multifaceted doers, as capable of rigging an irrigation system as managing a thousand transplants or moving a herd of goats.
>
> These characteristics in themselves add up to a lot of farm resilience. And that resilience

> explains how many farmers and their farms survive year after year.[24]

The Holmes family loved what they did and survived for 101 years though passion, resilience, and being multi-talented. As Foley points out, "Farmers are jacks of all trades, masters of multiple tools. They are also intimately tied to the resources that enable them to farm, from climate to soil to energy."[25] As the following pages reveal, the Holmeses personified these characteristics and ways of thinking. Foley lamented that "We have left behind much that farmers through the centuries learned and employed in their own struggles."[26]

Out of necessity, the Holmes farm was a self-sufficient enterprise, but not totally according to one recollection: "Most of their food was raised on the farm. They had to buy flour (unless they had wheat to have ground, which they sometimes did), molasses and salt. White sugar was a treat they seldom had."[27]

To keep afloat, the farm carried out activities to make money and sustain life on the farm.

## *Farm Revenue*

There were several sources of cash income.

### *Wheat*

Growing wheat was essential for the functioning of the farm. In the early years, the wheat was harvested with hand cradles, and later they bought a threshing machine. Wheat was a source of income. Unfortunately, selling wheat caused an accident that affected the history of the family. Nicholas's son, Jonathan, transported wheat by sailboat to Plattsburgh, New York, for sale. On one of the trips Jonathan's boat capsized near Valcour Island, and the entire load was lost. Worse, Jonathan never fully recovered from his immersion in the icy waters and suffered debilitating rheumatism until his death in 1894.

### *Chickens*

Raising chickens was a common and valuable undertaking on almost all farms, for the eggs, the meat, and cash sales. Jonathan's wife,

Hannah, made a business out of her chickens. She wrote this about what she called her "chicken business."

> You asked about the chicken trade. I have done better than any of my neighbors this winter so of course I feel well over it. Eggs have been scarce and high this winter. Charles and Alex have gone to the city (Burlington) today on the lake. I sent 24 dozen by him but don't expect a high price for them. I have gathered these in just three weeks besides what we have used to cook with. They are a dollar a dozen in the city of New York and I thought I should do better to go down there with them. The first of March shall begin to set them for hatching.
> (Feb 1881)
>
> I am glad to pursue my chicken business for it gives me a chance to get out of doors. I sold enough chickens to pay for a quarter of lessons for Lizzie so we will work to live as when you were with us. I shall know how many eggs I have brought in at the end of the year, also, when I have sold.[28]
> (Aug 1882)

*Sheep*

In the early 1800s sheep-raising was an important part of the Vermont economy, and the Holmes farm was active on this front. According to the 1826 Grand List, the Holmes farm had 26 sheep. Despite the decline in the Vermont sheep industry after the 1840s, the Holmes flock grew to 100 by 1858. As one observer wrote:

> In 1840, Vermont had 1,681,819 sheep, or nearly six sheep for every human. Mills for processing the clip were numerous in every county except in Essex and Orleans in the northeast. By 1850, sheep were statistically in decline in Vermont. By 1850

> dairying had grown as important as sheep raising in Vermont agriculture.[29]

It appears that the farm did not participate in the sheep "craze" that swept Vermont in the early 1800s. A leader of the movement was William Jarvis of Weathersfield, Vermont, who introduced 400 Spanish Merino sheep to his farm in 1811. The Merinos were a much-improved breed and had a significant impact on the Vermont and national sheep industry. The Holmes farm raised other breeds.[30]

The picture below (Figure 3.16), taken near the main barn around 1900, indicates that the Holmes sheep may have been Suffolk sheep, which arrived in New York State in the late 1800s. They are characterized by clean legs and heads, as displayed in the picture. Alternatively, because the picture shows that the sheep appear to be "speckle-faced," they may have been a cross between the Shropshire and Dorset breeds.[31] These breeds were bred for fiber and mutton and tended to be a supplement to other cash crops. Because of the substantial size of the Holmes flock by 1858, it is likely that sheep represented both a source of income (wool sold to a local mill) and a resource for operations at the farm, including manure, fiber, and food for the family.

*Figure 3.16 Sheep*

To accommodate the larger flock in the 1850s, the family needed more pasture space and, as noted above, purchased 40-acre Birch Island in 1859 for summer pasturing. Despite the receding sheep market, raising sheep continued to be a useful enterprise for the farm. Twenty-eight years later in 1887, they sold the island. We do not know whether this decision was driven by the fading market for Vermont sheep, altered priorities on the farm, or whether they needed the $2,000 in cash.

*Dairy cows*

Like almost all Vermont farmers, the farm had dairy cows. The Charlotte Grand List reported that the farm had 11 cows in 1858. Robert Holmes, who was born in 1885 and lived on the farm until 1923, recalled that they had 25 cows. It is likely that the family consumed milk from its cows along with making and consuming by-products such as butter and cheese. Also, with their sizeable herd, they delivered milk and butter to a local creamery, which created a modest income stream.

*Fruits*

The 1886 History of Chittenden County reported that the Holmes farm had 500 pear trees and 400 plum trees.[32] Did they sell this produce or keep it for their own consumption, or both? We don't know. If nothing else, these numbers suggest that the Holmeses did not do things in a small way. Even assuming a healthy harvest, that was a lot of pears and plums.

A hint that these fruits were a business venture is the way the farm was represented in the *Business Directory of Chittenden County, Vermont for 1882–83*. The entry mentioned "Charles Titus," Jonathan's son, as a "farmer," but showed C.T.s father as "Holmes, John, fruit grower."[33]

*Other revenue (not strictly agricultural)*

Three other revenue-producing enterprises—bricks, apples and horses—are described later. These "side businesses" were major commitments of the family and a source of recognition.

## *Other Farm Activities*

There were farming activities that were not, *per se,* a business enterprise. Rather, these were activities that supported the life of the family (e.g., food, ice) or facilitated the operations of the farm (e.g., manure, hay).

### *Hay*

Growing and harvesting hay was mother's milk for sustaining the animals on the farm. The summer cutting was a family undertaking, from which everyone took silent satisfaction: one could see the results (a shaved field) and savor the toil that made it happen. Figure 3.17 shows haying in progress at the farm.

*Figure 3.17 Haying*

Although the sale of hay was not usually a part of the Holmes enterprise (they used it for their own needs), their neighbors and relatives, the Whalley's, relied on this trade. A relative made the winter trek to Burlington to make a delivery:

> In the winter he drew hay from Charlotte to Burlington over the frozen lake using horses and

> sleds. It would be dark for the return trip so the horses had to find their own way around ices holes, etc. and always brought him home safely.[34]

*Potatoes and squash*

A small monograph published in 1976, called *Looking Around Charlotte,* mentioned the Holmes farm and pointed out that potatoes and squash "thrived on this extensive farm."[35] There is no evidence this produce went to market, so we can assume that potatoes and squash were grown primarily for family consumption. Excess produce was shared with nearby relatives and neighbors.

*Ice harvesting*

Keeping produce and meats safe for consumption over many months meant that the farm needed ice. They had an ice house on the southwest corner of the property and, with proximity to Lake Champlain and its annual icing over, used the winter months to harvest ice.

Beginning usually in late January when the ice was deep and solid enough, they employed the traditional method: cutting blocks with an ice saw with a huge blade mounted on a sled pulled by horses or oxen. Frigid and windy as any day might be in the winter, cutting ice was satisfying, sweaty exercise. Below, from a collection of old photos at the Charlotte Town Library is a picture (Figure 3.18) taken in 1922 of ice cutting at nearby Hill's Bay.

*Figure 3.18 Harvesting ice*

Storing ice in the icehouse on the farm had a prescribed method. In his book about living on the lake, Murray Hoyt described how it was done:

> The man who was filling his ice house would place the ice in it cake by cake, in a set pattern. When he had finished putting in the first layer he would start a second right where he started the first. He would continue with a third, a fourth, and as many more as required to fill the icehouse. He would pack sawdust in the space between the boards pf the building and each layer as he finished it, and would spread a thick layer of sawdust when he had finished placing the last.[36]

Like so many other farm operations, harvesting and storing ice had a methodology developed over years of trial and error.

*Manure*

Farming is a process of recycling. The farmer plants and grows hay, the hay is eaten by cows, cows produce manure, manure is spread on the hay fields, hay grows, and the cycle is repeated. A farm's manure, as harsh to the olfactory senses as it may be, is fundamental to the farming process.

Like, other farms, gathering and applying manure was an important task on the Holmes farm. In addition, the apple orchard had its own needs. C.T. discovered that intense manuring of the orchard paid significant dividends in the harvest. In the early 1900s, he used 250 tons of manure annually to cultivate the apple trees.

We don't know whether all the needed manure came solely from the farm or whether they had to purchase (or barter) from other farms. What we know is that manure was essential to making the farm a success.

*Oxen*

In the absence of tractors, there was always a need to pull and move things. Human muscle power was critical in countless chores but was supplemented by oxen and draft horses. Oxen were preeminent on Vermont farms through most of the 1800s and into the early 1940s.

An old Vermont farmer recalled his experience with oxen: "Yea, I am old ox teamster. Ever since I was a boy, I used steers. There's a lot of difference in teams, but they're much better to plow with than horses because they're steady. They take one gait and they'll follow it all day."[37] Town records from 1860 show that the Holmes farm had four oxen, with two working and two probably not yet grown to maturity.

## *Construction and Carpentry*

The Holmeses built three substantial houses and numerous out-buildings—large and small—on the property. This construction required knowledge and ability: building design, load factors, roofing, wood-cutting, brick laying, plumbing, etc.

These skill sets were passed down from generation to generation. I remember the passion and skill that my grandfather, Robert, who grew up on the farm, had for building things: a family summer camp on Lake Dunmore near Middlebury with a guest cottage; a trailer—

an RV—that slept three and made a cross-country roundtrip, a dog house, a wooden cross-bow, an intricate dollhouse, and a rolling cart for projecting color slides on a screen.

### *Inventory and Budget Management*

Maintaining and documenting a large inventory was vital to farm management, including land, buildings, animals, produce, machinery, and roads. The U.S. Agriculture Census of 1860 captured the holdings of the Holmes farm:

- Real estate value: $11,000; personal estate value: $5,000
- Town of Charlotte Grand List (Holmes and son): One poll tax $2; farm, 245 acres; $4,570
- Town of Charlotte: Total Value Real Estate, $4,570
- Total Amount of Charlotte Tax: $49.97
- 1860 Ag Census: 200 Improved Acres, 10 Unimproved, Value $11,000: 100 sheep, six horses, one colt, four oxen (two working), 11 cows, four swine, nine milk cows, 14 other cattle, 50 bushels wheat, 105 bushels peas and beans, 500 bushels oats, 400 bushels Indian corn, 200 bushels Irish potatoes, 116 bushels barley, 500 lbs. butter, 40 tons hay, 1,000 lbs. cheese (value of orchard produce—$100; value of farm machinery and implements—$500; value of livestock—$1,000)[38]

The farm consisted of multiple businesses and related activities and was a year-round enterprise that required the full energy and commitment of the entire Holmes family. It also required careful accounting of property and a family budget that tracked income flow, payments, ongoing expenses, and cash on hand.

## C. Bricks, Horses, and Apples

In addition to the farming enterprise, the family built substantial income-producing endeavors with bricks, horses, and apples.

### *The Brickyard*

As mentioned, the main house for Nicholas' family was built from bricks made from clay—a clay that was ideal for brickmaking—found near Holmes Creek on the northeast corner of the property. Lacking sophisticated machinery at the time Nicholas built his house, he used the classic, labor-intensive method.

Step 1: Dig up clay soil with a shovel to use as the material to build the bricks.

Step 2: Grind the clay soil into a fine powder or paste using large stones as "grinding stones." This will help in the bonding process. Grinding will generally improve the end result of the brick building.

Step 3: Mix the ground clay soil with water *to make a thick*, malleable paste. Traditionally, clay bricks were molded by hand or with wooden tools. Later, many clay bricks were made in purpose-built molds. Aim to produce clay bricks of the same size and shape.

Step 4: Leave the bricks out in the sun to dry or dry them in an oven to remove the moisture. Traditionally, brick builders lay bricks out in the sun to dry. An advantage of this method is that it does not require fuel or electricity.

Step 5: Fire the dried bricks in a kiln. A kiln is a type of ceramic oven that heats the contents to a very high temperature *for* several hours. This will alter the molecular bonds of clay and bind it to the other molecules, making solid bricks as a result.

In addition to meeting the construction needs of the property, Nicholas and his sons created a business out of brick-making. They made and sold the bricks for a large house on Mutton Hill in Charlotte, which became a stagecoach inn, and they produced bricks for several other houses in Charlotte and Burlington.

Nicholas started the brick operation early in his residence at the farm. When a house was demolished in Burlington in 1967, a brick was found that was inscribed: "John Holmes, Esq. 1834." This John was Nicholas's son, Jonathan, who was a young man of 14 at the time. We can assume that Jonathan had done work on the bricks.

The Holmes brick-making operation intersected with the construction project of a Charlotte church. Holmes bricks were used to rebuild the Charlotte Meeting House in 1840 after the previous structure was destroyed by fire. A Charlotte community newspaper in the 1960s reported this slice of history:

> The 36-foot by 48-foot structure replaced one previously destroyed by fire. It would appear that in the reconstruction some 1,000 or more bricks were donated by the Holmes family, who operated a local brickyard, toward the rebuilding project. Being the sharp Vermonters that they were, the elders of the church suggested to Mr. Holmes that perhaps he might wish to subscribe a pew as well. Since he happened to be a Quaker by persuasion, the gentleman felt that he had done his bit for this Methodist facility, and politely declined the suggestion, whereupon the matter was quietly dropped.[39]

A gift of 1,000 bricks was enough for Nicholas, especially since his first obligation, as a Quaker, was to the Friends. In 1950, after receiving severe damage from a windstorm, the building was donated to the Shelburne Museum. The local newspaper wrote:

> Neighbors well remember the meticulous care with which museum personnel carefully coded each

> brick as it removed and slid them down a cloth chute into the hands of loaders in the truck below to be transported to Shelburne.

Looking back, it is satisfying to know that the Holmes bricks were handled with respect. The reconstructed Charlotte Meeting House can be seen today on the grounds of the Shelburne Museum. The beautiful proportions of the building (Figure 3.19) are captured on the facing page.

*Figure 3.19 Charlotte Meeting House*

The available records do not indicate how profitable brick-building was to the bottom line of the farm or for how long the family built and sold bricks as a business. Today, vestiges of the old brickyard can be seen on certain days along the shoreline. The water is stained from the run-off from the clay that was ideal in making bricks beginning in the early 1800s.

## *Horses and the Morgan Heritage*

The Holmes family was passionate about horses. They rode them, raced them, bred them, and sold them.

During the 1870s, Jonathan and his son C.T. began to raise "blooded horses" for sale under the name, Apple Tree Farm. Their specialty was trotters, and they had as many as 75 horses on the farm. They also had a half-mile track near the stable where they trained the horses. The location of the track on the east side of the property is shown below in a 2020 photo (Figure 3.20).

*Figure 3.20 Site of farm racetrack © Emily Anderson*

Charles Ross, whose mother, Hannah Elizabeth, grew up on the farm, was proud of the Holmes heritage in raising Morgan horses. In an interview for an oral history project, Charlie pointed out that, "Part of the tradition of our family was that they were among the largest, most respected breeders of horses in the late 1800s. So, horses were a part of my heritage."[40]

> C.T., my grandfather on my mother's side, had been into horses since about 1850. From then through about 1915, they raised and trained horses as a sideline. They were trying to cross Hambletonian Messenger, and a number of other extremely well-known thoroughbred stallions to an offspring of Justin Morgan.
>
> The Morgan horse is so inextricably tied to Vermont life that it is almost impossible to overlook its role. Moreover, in my case, my mother's family, aside from being farmers and orchardists, were also engaged in the raising and training of horses for more than 75 years. The strength, the beauty, and the gentleness of the Morgan horse is an inspiration to every real Vermonter.[41]

Charlie went on to tell how Justin Morgan "Brought his young colt to Vermont and the colt became famous because he was so prepotent. He was able to transmit his characteristics to his offspring."[42] In time, the horse took on the name of its owner.

> He had several famous sons who were bred extensively in the early 1800s. The offspring were valuable in Vermont because they were useful for fieldwork, for saddle riding, and as driving horses. As it turned out, one of Justin Morgan's descendants, Ethan Allen 50, was the fastest racehorse pulling a cart in the United States. He was raised in the Champlain Valley and stood at stud throughout the area. My mother's family had a granddaughter by Ethan Allen 50, which was one of the founders of a cross they achieved.[43]

The Holmes family developed a stable of horses with the Ethan Allen 50 bloodline and entered the trotting world. They "pulled a cart," to use Charlie's phrase. In the picture below (Figure 3.21), C.T.'s son, Robert, shows off a Holmes sulky.

*Figure 3.21 Robert with sulky*

While the they were breeding horses with the Morgan bloodline, Middlebury's Joseph Battell took actions to ensure that the Morgan horse had a prominent—and permanent—presence in Vermont and beyond. According to a *Vermont History* essay, "Battell began breeding horses in 1875 and bred nearly 150 of them during the next 40 years of his life."[44]

Battell was instrumental in promoting the assets of the Morgan horse among horse people across the nation. In 1878 Battell built a magnificent barn for his horses on land he had purchased in Weybridge, outside of Middlebury. At age 68, in 1907, he deeded the land and building to the U.S. Government, and the horses bred and trained at the farm were used extensively by the U.S. Calvary until World War II. Sadly, after it became clear that horses were no longer vital to warfare, many horses were destroyed.

The remaining government horses were dispersed to the New England land grant universities, except for 26 Morgans that remained at the farm. In 1951 the farm was taken over by the University of Vermont and became known as the UVM Morgan Horse Farm.[45] Ever since, Morgan lovers from across the nation, including the extended Holmes family, have made the pilgrimage to Weybridge.

Battell's horse farm had an impact on young Charlie Ross, who grew up in Middlebury and developed a deep affection for the Morgan horse. A regular outing for Charlie and his family was a visit to Battell's farm in Weybridge to admire the Morgans. In addition, his father, Dr. Ross, used a Morgan to pull his carriage on house calls.

After Charlie completed service on the Federal Power Commission in the Kennedy and Johnson administrations, he returned to Vermont and bought property in Hinesburg. He started a Morgan horse farm, Taproot, where the family bred Morgan horses. His daughter, Jacqueline, showed Morgans throughout the East Coast, and Taproot's Morgans, and Jackie, developed a national reputation.

As for the Holmes farm, the development of the horse business culminated in the publishing of their first sales catalogue in 1885 (Figure 3.22).[46]

CATALOGUE

OF

THE TROTTING STOCK

AT

APPLE TREE FARM.

HOLMES BROTHERS.

CHARLOTTE, VERMONT,
1885.

*Figure 3.22 Horse catalogue*

The pedigree of Champlain (Figure 3.23) includes this sale pitch: "Champlain is a colt of great beauty, good size, and shows wonderful natural speed, and should not only make a trotter but get them, as he combines the blood of Mambrian, Tysdyk's Hamiltonian, Black Hawk, and Bashaw."

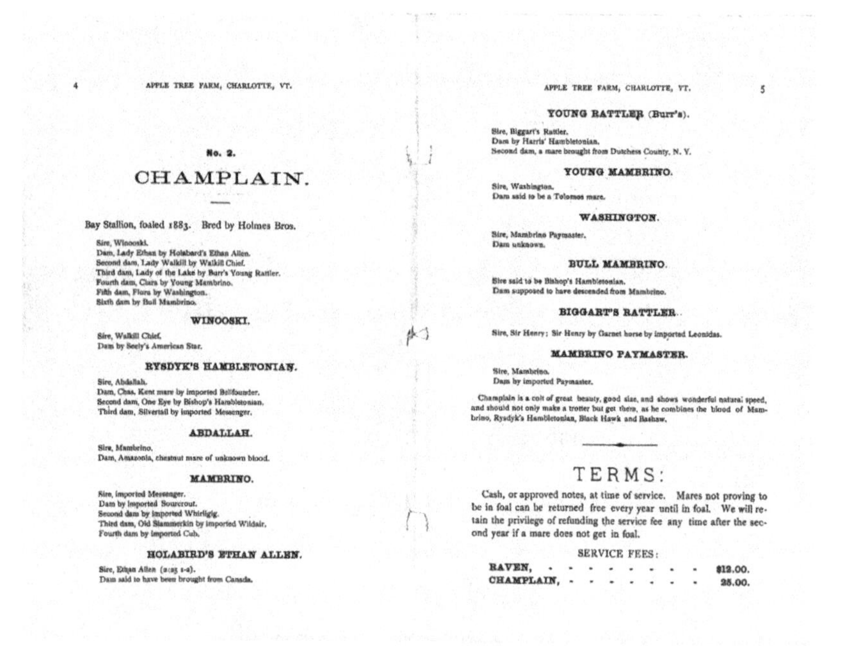

4 APPLE TREE FARM, CHARLOTTE, VT.

No. 2.

CHAMPLAIN.

Bay Stallion, foaled 1883. Bred by Holmes Bros.

Sire, Winooski.
Dam, Lady Ethan by Holabard's Ethan Allen.
Second dam, Lady Walkill by Walkill Chief.
Third dam, Lady of the Lake by Burr's Young Rattler.
Fourth dam, Clara by Young Mambrino.
Fifth dam, Flora by Washington.
Sixth dam by Bull Mambrino.

WINOOSKI.

Sire, Walkill Chief.
Dam by Seely's American Star.

RYSDYK'S HAMBLETONIAN.

Sire, Abdallah.
Dam, Chas. Kent mare by imported Bellfounder.
Second dam, One Eye by Bishop's Hambletonian.
Third dam, Silvertail by imported Messenger.

ABDALLAH.

Sire, Mambrino.
Dam, Amazonia, chestnut mare of unknown blood.

MAMBRINO.

Sire, imported Messenger.
Dam by imported Sourcrout.
Second dam by imported Whirligig.
Third dam, Old Slammerkin by imported Wildair.
Fourth dam by imported Cub.

HOLABIRD'S ETHAN ALLEN.

Sire, Ethan Allen (2:25 1-4).
Dam said to have been brought from Canada.

APPLE TREE FARM, CHARLOTTE, VT. 5

YOUNG RATTLER (Burr's).

Sire, Biggart's Rattler.
Dam by Harris' Hambletonian.
Second dam, a mare brought from Dutchess County, N. Y.

YOUNG MAMBRINO.

Sire, Washington.
Dam said to be a Tolomos mare.

WASHINGTON.

Sire, Mambrino Paymaster.
Dam unknown.

BULL MAMBRINO.

Sire said to be Bishop's Hambletonian.
Dam supposed to have descended from Mambrino.

BIGGART'S RATTLER.

Sire, Sir Henry; Sir Henry by Garnet horse by imported Leonidas.

MAMBRINO PAYMASTER.

Sire, Mambrino.
Dam by imported Paymaster.

Champlain is a colt of great beauty, good size, and shows wonderful natural speed, and should not only make a trotter but get them, as he combines the blood of Mambrino, Rysdyk's Hambletonian, Black Hawk and Bashaw.

TERMS:

Cash, or approved notes, at time of service. Mares not proving to be in foal can be returned free every year until in foal. We will retain the privilege of refunding the service fee any time after the second year if a mare does not get in foal.

SERVICE FEES:

RAVEN, - - - - - - - - - $12.00.
CHAMPLAIN, - - - - - - - - 25.00.

*Figure 3.23 Champlain lines*

Sales were as far afield as South America, including the sale of their best stallion, "Nero," in 1887. They also sold horses at Madison Square Garden in New York City.

Because their father Jonathan was almost completely incapacitated by this time, his sons C.T. and William were the named sponsors of the sale. Ten horses were presented, with the catalogue noting "we simply wish to call attention to the rare combination of trotting blood which embraces the families of Mambrino, Seely's American Star, Rysdyk's Hambletonian, Black Hawk, Smuggler, and Bashaw." The professionalism of the presentation was captured in each horse's bloodlines.

*Racing horses*

C.T. and his son Robert trained their horses on the half-mile track on the Holmes farm. Figure 3.24 shows Robert pulling a cart.

*Figure 3.24 Robert on track*

C.T. raced trotters and pacers on the fair circuit from Canada to Rhode Island in a high-wheeled sulky. Typically, the drivers raced for best horse in three-of-five heats for purses of up to $300.

C.T., his son Robert, and other relatives raced on the two local tracks in Charlotte. Russell Williams, a Charlotte resident and member of a longtime local family, described the local racing scene in a work published in 1980 by the Charlotte Historical Society:

> For years and years everyone owned a horse which was believed to be faster than any other horse in the area. The urge to prove the superior speed of one horse over another persisted, so, in East Charlotte, back of the St. George place, a trotting track was made. There during evenings and Saturday afternoons, people could hitch up a sulky and give the horse his (her) head.
>
> For some reason, after a year or two, a track was built in the western part of town and a regular schedule of weekend races was run, bringing in drivers from neighboring towns to enter their

> horses in competition. The races drew good crowds. the ladies of the Breezy Point Library Association, always eager to raise funds for their cause, sold soft drinks. Gilman Foote, on the lively black horse named Jim, was the Grand Marshall and kept everything running smoothly. Sometimes the horses didn't understand how everything should work at the start. In these days there were no starting gates, and unless the horses were evenly abreast at the starters signal, they would all be called back. This often went on for several tries, but since it was part of horse racing, the crowd didn't get too restive.

Williams pointed out that the only prize for the winning horse was a "good wool blanket to curl up with on chilly nights." Just north of the Holmes farm on Hill's Bay, he said, winter trotting races were held on the ice.

> For various reasons they did not do as well as the summer races. The ice was uncertain, the horses had to be shod with ice corks (calks), and runners substituted for wheels. And, it was cold.[47]

The Holmes's were in the middle of the Charlotte race scene and, as wild and unruly as it was, were sufficiently successful that they raced their horses on the regional racing circuit.

On the facing page (Figure 3.25) is an announcement for local races in 1923.[48] Although the Holmes family had moved away earlier that year, Robert and his brother-in-law, Charlie Johns, continued to race in Charlotte.

HORSE RACES

RECREATION PARK,

CHARLOTTE, VT.

Down by the Station

SATURDAY, AUG. 4, '23

Auspices of Charlotte Trotting Association

4 BIG CLASSES

PONY RACE DISPLAY OF STOCK

STATE SCHOOL BAND IN ATTENDANCE

RACES BEGIN AT 1:00 P. M. SHARP

Admission 35c, Autos and Children Free!

*Figure 3.25 Charlotte race broadsheet*

Race day (Figure 3.26) was a big event in Charlotte, with large turnouts of fans from Charlotte and nearby towns.

*Figure 3.26 Charlotte racetrack*

The picture below (Figure 3.27) shows Robert at one of the Charlotte tracks on race day.

*Figure 3.27 Robert at racetrack*

Robert and his brother-in-law, Charles Johns, raced the circuit, including events at the track in Saratoga. In an interview in 1958, Robert said race times advanced tremendously after the introduction of new rubber-tired, steel-spring sulkies. Still, as trotting began to take a backseat to thoroughbred racing at many tracks, Robert expressed his preference: "Sulky racing is the best, and jockey racing is not nearly the thrill."[49]

## *Apple Orchard*

By the late 1800s, the apple orchard became a major focus for the Holmes family and was an important source of income. It received wide notice in Vermont and beyond. The history of the orchard is described in Chapter Five. In addition, there were innovative offshoots of the orchard that brought income.

As noted, Vermont farms produced creative solutions to day-to-day challenges. The experimental and entrepreneurial attributes of C.T. are seen in two related ventures pertaining to fruit growing. The Official Gazette edition of Oct. 17, 1905, reported a "patent under Charles T. Holmes filed Dec. 14, 1904 for a pruning saw." (Figure

3.28)[50] We don't know whether C.T. ever produced the pruning tool and brought it to market.

802,290. PRUNING-SAW. CHARLES T. HOLMES, Charlotte, Vt. Filed Dec. 14, 1904. Serial No. 236,873.

*Claim.*—1. A pruning device, consisting of a handle, outwardly-extending plates carried by the handle and having oppositely-arranged openings, a saw-blade pivotally mounted between said plates and having a series of segmentally-arranged openings, and a spring-arm carried by the handle and having an inwardly-extending stud passing through the opening in one of said plates and one of the series of openings in the blade and entering the opening in the opposite plate.

2. A pruning device, consisting of a handle, outwardly-extending plates carried by the handle, a saw-blade pivotally mounted between said plates and having a series of openings on the inside of the pivotal connection, and a spring-arm carried by the handle and having an inwardly-extending stud passing through an opening in one of said arms and adapted to enter one of said series of openings in the saw-blade.

*Figure 3.28 Pruning saw patent*

At about the same time, an advertisement (Figure 3.29) appeared in *Walton's Vermont Register* showing that C.T. was a sales representative for spray pumps and spraying materials.[51]

# SPRAY PUMPS

**Both hand and power. Order early and have them in time, as delays are dangerous. None better than those made by**

## Field Force Pump Co.

**of Elmira, N. Y. Strictly high pressure and prices right.**

**The only hand pump that has an Automatic Agitator.**

**Also look well to your spraying material. I have had the best results with Grasselli's Arsenate of Lead and all other products of their works made for spraying. Address or call on**

**C. T. HOLMES**

**Lake View Orchards CHARLOTTE, VT.**

*Figure 3.29 Spray pump advertisement*

C.T. was a well-known innovator in the use of spray pumps in orchards, so we can presume that he was well-positioned as a salesman. In 1919, for example, C.T. traveled to Middlebury to deliver one of his pumps to a customer.

## *Elements of Farming*

An often-unsung dimension of farming is the extent of knowledge and expertise required. The varied farming activities, the successful brickyard, the thriving orchard, and the raising, training, and racing of trotters demanded authentic expertise on the part of the Holmes family.

This knowledge and expertise arose from word-of-mouth, hands-on training among relatives and acquaintances, trial and error in the face of challenges, interacting with nearby experts (e.g., Vermont Department of Agriculture; apple people; horse people), participation in professional organizations (e.g., Vermont State Horticultural Society), and research when available (books and journals).

In a sense, the farm was its own "university." It provided both a general education (learning across many fields) and a technical education (e.g., knowing how to spray an apple orchard for insects or how to breed horses). Absent a college education, these were well-informed, intelligent people who knew how to do things well.

Whereas the Holmes farm was a vibrant, multi-dimensional enterprise from the 1800s into the early 1900s, the period after the Civil War has been the subject of debate among historians. A view that prevailed for a half-century into the 1980s was that this was an era of stagnation and deterioration in Vermont: Loss of hill farms, lacking population growth, out-migration, and dying villages.

In a seminal essay in 1984, however, Nick Muller took a fresh look at the 1870–1900 period. In what he called "A New Look at the Neglected Winter of Vermont," Muller suggested "the need to replace the prevailing view articulated by Harold F. Wilson's metaphor of a 'severe winter season.'"[52] He wrote:

> The migration from Vermont, a central element in the thesis of a moribund, declining society, which

> did retard the growth in total population, did not necessarily strip the state of its most able and energetic citizens. These were not quiet, depressed years in which a previously rambunctious society lost it vitality in the face of migration and difficult economic circumstances.[53]

The Holmes farm exemplified the vital, entrepreneurial side of Vermont, and the winter metaphor is not an accurate representation. Yet, it is evident that the time after the Civil War was anything but smooth for farmers and, as Kevin Graffagnino pointed out, "Relatively few Vermonters prospered at sheep or dairy farming during the railroad era."[54]

After the Civil War, most Green Mountain farmers found it necessary to diversify into the production of maple sugar, apples, firewood, potatoes, corn, wheat, Morgan horses, and a variety of other crops and livestock just to make a living.[55] As shown in this chapter, the Holmes farm was a prime example of diversification. The aggressive push into apples and horses was the most telling example.

Farming has an eternal rhythm. When one looks at Vermont farms of today, not much has changed. Although many contemporary farms operate on a far larger scale (size tends to be a safety net against financial volatility) and use a wide array of expensive machinery, farming work never ends. A farmer's year is a constant—a rhythm that repeats itself into the future. Vermonters know what it takes to farm, and, except for the usual aggravations ("Will the rain ever stop?") and the yearly unknowns about prices ("You mean, they are paying *that* for milk?"), Vermont farmers find deep satisfaction in the products of their work: The hay brought in, the manure spread, the corn "knee high by the 4th of July," pies made from homegrown apples, and much more.

It is who they are. It was who the Holmeses were.

## Chapter Four

# Life on the Farm

*"Grandmother needed her sons' help on the farm, so when the sons married, each brought his bride to live in the house, the three families living together in the large brick house on the shores of Lake Champlain."*

With multiple enterprises, a large inventory of machinery and buildings, unknowns about each year's harvest, complex bookkeeping, and more than one generation living and working on the farm, there was inherent complexity in managing the farm and in daily life. This leads to fundamental questions about life on the Holmes farm.

We know that family members were in close and continuous contact as they engaged in a multi-faceted business with no guarantee of success. From a distant vantage point of more than 100 years, it is difficult to discern with confidence the social structure of the farm. What was the division of a labor? Who had influence, and who was perceived to be the "boss"?

Looking deeper, what was the character of the people? Among the members of this hard-working, strong-willed family, did harmony prevail? How did they relate to each other? Were people competitive and manipulative? Was there a core foundation of love, concern and commitment?

For sure, there was strength of character. There was resilience, perseverance, and grit. But, looking deeper, what was it like to live on the farm? If one was sent back in time to the farm, how did it feel to be there? We get useful clues from what we know about the people—their schooling, their get-togethers, their forms of recreation, their religion, and their passions.

## A. The People

The Holmes farm existed as a community unto itself, a small hamlet of multiple buildings, three generations of relatives and, during apple-picking time, many hired hands. The first clue to life on the farm is what family members—and how many—lived on the farm over 101 years. Beginning with Nicholas' move from Monkton in 1822 with his wife and children, here are the households over four generations, with the birthdates of spouses and children.

First Generation

*Nicholas Holmes (born 1782)*
*Wife: Sarah Hazard (1791)*
*Children: Robert (1813)*
*Nicholas (1815)*
*Jonathan (1820)*
(Born on the farm)
*Mary (1824)*
*Julia Ann (1826)*

Second Generation

*Jonathan, son of Nicholas (Born 1820)*
*Wife: Hannah Smith (1822)*
*Children: Gertrude (1847)*
*William Hazard (1852)*
*Charles Titus (1857)*
*Mary (1858)*
*Elizabeth or "Lizzie" (1861)*

Third Generation

*William Hazard, son of Jonathan (born 1852)*
*Wife: Mary Ann Sherman (1853)*
*Children: William Sherman (1877)*
*Edith (1879)*
*John (1880)*
*Carleton (1883)*
*Harriet (1885)*
*Alice (1891)*
*Alfred (1893)*

*Charles Titus ("C.T."), son of Jonathan (born 1857)*
*Wife: Clara Russell (1861)*
*Children: Hannah Elizabeth (1881)*
*Maude (1890)*
*Mildred (1889)*
*Robert (1885)*

Fourth Generation

*Robert Holmes, son of Charles Titus (born 1885)*
*Wife: Rena Johns (1882)*
*Children: Marion (1912)*
*John (1915)*

Over the life of the farm, 31 family members lived on it. Among the heads of the farm, Nicholas (1782–1863) lived to 81, Jonathan (1820–1894) lived to 74, William (1852–1929) lived to 77, and C.T. (1857–1930) lived to 73. C.T.' s son, Robert (1885–1968), lived to 83. Twenty children were born on the farm or nearby.

Whereas CT's daughter, Mildred, died at three years of age in 1892, there is no record of infant deaths. It is possible, however, that an infant death may have gone unrecorded. It was the practice in some families to not record an infant death and, on occasion, name a succeeding child by the same name. In fact, town records show a Mildred Russell Holmes was born in 1897 and died of convulsions five years later. Thus, it is possible that there was another Holmes child who died young.

Among the male children, Nicholas' son, also Nicholas, was older than his brother, Jonathan, but Jonathan succeeded his father in the leadership of the farm. Similarly, Jonathan's son, William Hazard Holmes, was older than C.T. but C.T. succeeded Jonathan. We don't know the reasons for the ascendance of Jonathan and C.T. over their older siblings.

Living together at the farm with overlapping generations was an experience that is rare in today's world. Saved in the archives of the farm is a humorous poem entitled "A Little Boy's Lament," which appeared in a local newspaper. The poem by A.J. Worden expresses a boy's affection for his grandfather and the freedom he has away from the oversight of his parents. Here is an excerpt:

*I'm going back now down to grandpa's,*
*I won't come back no more*
*To hear the remarks about my feet*
*A muddyin' up the floor.*
*They's too much said about my clothes,*
*The scoldin's never done —*
*I'm goin' back down to grandpa's,*
*Where a boy kin hev some fun.*
*I bet you grandpa's lonesome,*
*I don't care what you say;*
*I see him kinder cryin'*
*When you took me away*
*When you talk to me of heaven*
*Where all the good folks go,*
*I guess I'll go to Grandpa's,*
*An' we'll have good times, I know.*[1]

The poem's theme returns me to memories of my grandfather Holmes, who grew up on the farm. Robert took me camping, drove me in his truck on his sales route with the Central Vermont Public Service Corporation, and invented things for me, including a wooden crossbow that shot arrows about 30 feet. In the phrase of the psychologist, Carl Rogers, I sensed my grandfather's "unconditional positive regard."

There are scarce documents describing Nicholas or his wife, Sarah. In letters written in 1880s and 1890s, we know more about his son, Jonathan, his wife, Hannah, and their children.

## *Jonathan (1820–1894)*

Jonathan grew up on the farm and went to local schools, including his early years at a school about a mile from the farm. It appears that Jonathan's school, at least for a winter term, was one of great harmony. When Jonathan was 11, he received a letter—penned on March 8, 1831—from his schoolmaster that noted that "the period of our separation having arrived." Here is one paragraph in a long letter of good-bye:

> When I reflect upon the almost uninterrupted concord which has prevailed in our little community, upon the general condescension which has marked your intercourse with each other, and upon the general care you have evinced, to facilitate, rather than retard the operation of the various plans I have adopted for the improvement and welfare of the school—and when in contrast therewith, I reflect upon those scenes of disorder, which so frequently occur in our common schools, upon the obstinacy of the scholars, the petulance, the threats, and the punishments of the scholars, and mutual dislike of all, I cannot well forbear to indulge those emotions of delight, which the recollection of our harmonious proceedings and of your progress in study, sends to inspire.[2]

Was a 11-year-old advanced enough to penetrate the run-on sentences and florid language of the teacher? Since there was a mixture of levels and ages in the small school, the letter was probably directed to the older students and their parents. Yet, if this small school not far from the Holmes farm was uniquely harmonious, perhaps the Quaker upbringing and respectful style of the students explains the apparent good harmony.

With schooling in the basic academic subjects and exposure to literature, but not much more, Jonathan went on to be the patriarch of the family and farm. Below is a photograph (Figure 4.1) of Jonathan taken around 1850 at the age of 30. Although photographs can be deceiving, Jonathan's eyes suggest a man of seriousness and humility.

*Figure 4.1 Jonathan at 40*

His father, Nicholas, died in 1863 when Jonathan was 43, but he already was a close partner with Nicholas on farm matters.

Unfortunately, just 15 years later around 1877, Jonathan spilled into the icy waters of Lake Champlain while transporting wheat to New York State. He soon developed debilitating arthritis and never recovered. Jonathan was the subject of great concern after his fall into the lake because of his deteriorating condition, and we have this testimony about Jonathan's illness and its effect on the family:

> The wheat which was raised on the farm was transported by sail boat to Plattsburgh and, on one trip, John capsized off Valcour Island losing the entire cargo. To his immersion in the icy waters was attributed the fact that he was bedridden for many years at the end of life with a severe rheumatic condition.[3]

> John Holmes, although seriously affected by a combination of rheumatic difficulties, is ever cheerful.[4]

> Grandfather was an invalid for years, crippling arthritis, but called rheumatism then.
> *–Jonathan's granddaughter, Alice Holmes, 1881.*

> Christmas Day went off as gay as could be. We did get Father's consent to bundle him up and take him with us and we had a jolly time getting started for we shut the house up and Polly went too. Charlie drove Bill and Kentucky and the bob-sleds and we had a long box and all piled in. There were three seats and father in his chair rolled in last and Will stood behind him, so you see we had a driver and footman now and were stylish now.
> *–Jonathan's wife, Hannah in 1880 on stationary of the "APPLETREE STOCK FARM, J. Holmes & Sons, Proprietors."*

> Lizzie has not been home yet and we do think it is time she came to see her poor suffering father.
> *– Hannah in 1881*

> My poor John is unable to walk. We got him a new chair last winter which is a source of comfort for him.
> *– Hannah in 1882*

> The boys hired a professional nurse that can lift him.
> *– Jonathan's daughter-in-law, Mary Will, 1883*

The early 1890s were sad and debilitating years for Hannah and Jonathan.

> Father Holmes is perfectly helpless, has to be fed even. His mind is perfectly active. He reads a great deal but someone has to turn the pages or the leaves of his book as has no use of his hands. He remembers everything and converses on the topics of the day and enjoys a hearty laugh with an old associate as well as ever; of the two I think Mother quite as liable to wear out first. She has given up the entire care of the house to Clara (C.T.'s wife) save her room and what was the girls room and is still Lizzie's. She keeps the care of the hens as that takes her out in the open air.
> *– Mary Will, c. 1890*

> Father grows more helpless. He can open his mouth to take food only and it is difficult to understand his speech.
> *– Mary Will, 1892 or 1893*

> The past year has been a very sad one for us. Father Holmes, after years of continuous suffering, passed away February 18, 1894.
> – *Mary Will, January 1895*[5]

Jonathan lived through most of the 19th century. Following in his father's footsteps, he improved and grew the farm. Importantly for the farm's future, he set the apple orchard around 1862. Despite his enduring illness, he remained intellectually alive, kept up with current events, and loved to engage in conversation. Below (Figure 4.2) is an oil portrait of the full-bearded Jonathan which has been handed down through the years.

*Figure 4.2 Jonathan oil portrait*

## *Hannah (1822–1901)*

We know more of Jonathan's wife, Hannah. In addition to the above references, we have extensive testimony, beginning with Hannah's Quaker identity.

> I can tell you a bit about my grandmother Hannah Smith Holmes. My grandparents on my father's side (William Holmes) were Quakers. Grandmother (Hannah) was dressed in gray, wore her hair combed simply with two long curls or ringlets down in front of her ears, as was the Quaker style. In speaking she always used the Biblical thee, thou, thy form of pronouns.
> *–Alice Holmes, her granddaughter, in 1971*

> As a very young child, I liked to sit beside her at meal-time because her false teeth (dentures nowadays) made a fascinating clicking sound when she ate. They couldn't have been a very good fit, but even so, probably much superior to the wooden ones of George Washington's time! Isn't that a silly thing to stay in my mind all these years. I also remember the NECCO peppermint wafers she kept in her top bureau drawer which she would dole out to us grandchildren when we visited her.
> *–Alice, 1971*

> There was no Quaker Meeting House in Charlotte, the nearest one being in Ferrisburgh. My mother's family, the Sherman's of East Charlotte, attended services at the Congregational church (in Charlotte) that was where we seven children and Mother (Mary) went. Father (William) did not attend as Grandmother (Hannah) felt that church

> was too "worldly," and disapproved of Father's going.
> –*Alice, 1971*

Hannah's business acumen and sense of command were front and center.

> Grandmother was a very strong-minded person who ruled her household with a firm hand. Always just and always calm. She needed her sons' help on the farm, so when the two sons (Williams and Charles Titus) married, each brought his bride to live in the home, the three families living there together in the large brick house (now torn down) on the shores of Lake Champlain, and Grandmother controlled and managed the finances for all three families. I can imagine that arrangement may have presented some problems at times, especially after grandchildren began to arrive.
> –*Alice, 1971*

> By putting pieces of new carpet together, which took two weeks . . . now you would hardly know us we are so spruce . . . Clara (her son's wife) and I thought they were very nice. Just then Will (Hannah's son, 29 years old) came in after commenting on the good looks of things in general said 'that woman made a mistake, your carpet stripes don't match and he pointed to our pretty rugs, so we thought if they were not going to be appreciated better that we would put something down that would be so. I had cut some good pieces out of the old carpet and the next time I find anyone dares to say our carpet isn't matched I'll put them down and then see if they can tell the difference.
> – *Hannah on making a carpet, 1881*

> You asked about the chicken trade. I have done better than any of my neighbors this winter, so of course I feel well over it. Eggs have been scarce & high this winter. Charles & Alex have gone to the city (Burlington) today on the lake. I sent 24 dozen by him but don't expect a high price for them. I have gathered these in just three weeks besides what we have used to cook with. They are a dollar a dozen in the city of NY & I thought I should do better to go down there with them. The first of March shall begin to set them for hatching.
> *– Hannah on her chicken trade, Feb 1881*

> I am glad to pursue my chicken business for it gives me a chance to get out of doors. I sold enough chickens to pay for a quarter of lessons for Lizzie so we will work to live as when you were with us. I shall know how many eggs I have brought in at the end of the year, also, when I have sold.
> *– Hannah, 1882*

Hannah's last years were hard ones. She died seven years after Jonathan's passing.

> As each year passes me by, I find I am failing in strength to do what lays next to my hands.
> *– Hannah, 1882*

> Mother (Hannah) has given up much more of her work than when you were here . . . Mother has been very poorly all summer, not able to ride at all.
> *– Mary Will, 1882*

> Mother Holmes (Hannah) was confined to her bed about six weeks. She was completely worn

> out, is able to go about now but does not care for Father.
> *– Mary Will, 1883*

> Mother (Hannah) taken sick the week following (Jonathan's death) did not leave the house again until August. She is now quite well but her growing deafness is a great trial and she goes away from home very little.
> *– Mary Will, 1895*

When Hannah married Jonathan, she left the Smith family for the Holmes farm. She kept informed about the family, including her wayward brother, and earned sympathy from her daughter-in-law, Mary Will.

> I was also glad to hear that brother Thomas (Smith) was married, if it will only prove to be a happy thing for both parties, but I doubt and fear. I have felt so sad that a brother of mine should be a homeless wanderer, still we all know it is his own misdoing that make him so. If Eva can write anything that will make me feel better over his condition, I will take it kindly.
> *– Hannah, 1881*

> Mother Holmes (Hannah) is now the only one living of your Grandfather Smith's family . . . there must be a feeling of loneliness when the last parent has passed from earth.
> *– Mary Will, 1895*

The author of many of these letters—treasures of the past—was Mary Will Holmes, Jonathan and Hannah's daughter-in-law. Her letters affirm that Hannah was a sharp businesswoman, including overseeing her small chicken business and managing the affairs of the household.

With three families—Jonathan's, William's and C.T.'s—living together in the brick house, this was no small achievement. Hannah held firmly to her Quaker upbringing, dressed in a traditional way, and disapproved of the too-worldly Congregational Church of Charlotte. And she had the years-long burden of overseeing the care of her invalid husband. Yet, for all the challenges and work, she carried forth with a love of nature and an appreciation of literature, as shown in a letter to her sister:

> We are having such cool bright days. I have not forgotten those pretty lines "Oh, Gift of God, Oh, Perfect Day" that you and I used to repeat when the day would apply. I found after the death of my favorite poet "Longfellow" they were written by him. So many little children's hearts were sad over his death. His love of children, and truthful simplicity attracted them to him.
> – *Hannah, 1882*

Hannah, in her quiet and consistent way, was the backbone of the family for a long span of time. In the face of her husband's chronic illness and other challenges of the farming life, she kept things together for the family in a heroic way.

## *Jonathan and Hannah's Children*

Jonathan and Hannah had five children: three daughters, Gertrude, Mary, and Elizabeth, and two sons, William and Charles/C.T.

The boys stayed on the farm. Gertrude, the oldest, was born in 1847. She married and moved away. Mary, who was born in 1858, married a nearby Whalley and stayed close to the Holmes family throughout her life. Mary's daughter-in-law remembered Mary's stories of growing up on the farm:

> She (Mary) grew up at the Holmes' place with her brothers and sisters. Her oldest sister, Gertrude, used to drive the horses on the mowing machine when they first had a horse-drawn mower, to free the men and boys for heavier hand work. The children took lunches to the men in the fields cutting grain with cradles, five men following each other around the field. They had five meals a day at that time. The three regular meals, early breakfast, dinner, and late supper, with another morning and afternoon meal served in the fields.

At one point, Mary's parents went to St. Louis to visit her older sister Gertrude, who had married a cousin.

> They were gone several months in the winter and the children at home were cared for by Polly (the family retainer) who kept them fed and clothed, knitting socks, etc. and making clothes for them as needed. The boys were old enough to look after outside chores, milking the cows, feeding livestock, cutting wood, etc. [6]

We know little about Polly but it is noteworthy that the Holmes parents could afford help and trusted Polly to oversee the children in their absence.

### *Elizabeth Holmes (1861–1933)*

The youngest child of Jonathan and Hannah, Elizabeth, was known as Lizzie in her younger years. Lizzie pursued her dream of a career in music and married into a famous family. Hannah's letters tell about her trajectory into the music world, including Lizzie's study at a conservatory in St. Louis.

> Lizzie has gone, and we miss her so much. You know the baby in the family Is always missed

more than any other when they are gone. Mother (Hannah's mother, a Smith), Charlie (her son) and Clara (Charlie's wife) carried her up last Monday and we have not heard from her yet but expect to get a good report of her smartness tomorrow.
*–January 1881*

Lizzie has not been home yet and we do think it is time she came to see her poor suffering Father.
*– February 1881*

Lizzie is still in St. Louis at the Conservatory. The professor thinks she will not be obliged to stay two years longer and perhaps one and a half. I sold enough chickens to pay for a quarter of lessons for Lizzie.
*–August 1882*

Elizabeth has a position teaching vocal music in the public schools in Port Henry (Port Henry is across Lake Champlain in New York State, a few miles southwest of the Holmes farm). She has dropped her nickname Lizzie, and is known in the music world as Elizabeth.
*– November1882 or 1883*

It is not clear why Lizzie came home (across the lake from the farm) when it appeared she would be in St. Louis for another two years. In all likelihood, she could no longer afford the training. Nonetheless, while teaching music in Port Henry, Lizzie's work was well-received. A local news story, a clipping that was saved over the years by relatives, says:

The entertainment given by the scholars of the Port Henry Union School and the Choral Club under the direction of Miss Elizabeth Holmes at Lewald

> Opera House on Friday evening, May 5th was a success in every respect and fully deserved hearty applause which each number received, and Miss Holmes deserves great credit for her pains taking efforts with the children.[7]

With a teacher trained at prestigious conservatory in St. Louis and an opera house as a venue, it is noteworthy that the concert was performed by young children, grades 1–5. In addition to songs such as John Brown's "Ten Little Indians," many of the songs were patriotic in theme, including "flag" songs. The writer noted, "It was fully demonstrated how thoroughly each child entered into the soul stirring spirit of the occasion."

Soon thereafter, Lizzie moved to Boston.

> Lizzie is in Boston completing her musical studies. She makes her home at Mr. Monroe's whose son Kirk is an author. Perhaps you have seen some of his articles in Scribner's Magazine.
> *– circa 1883*

During Lizzie's sojourn in Boston in the 1880s to pursue her music career, she met Charles Stowe, son of Harriet Beecher Stowe. Harriet was author of the wildly successful book, *Uncle Tom's Cabin*, which depicted the plight of slaves and contributed to the abolitionist movement. Charles Stowe was educated in the literary and religious tradition of his family.

While Lizzie was studying in Boston, she boarded in Cambridge in the Monroe home and met Susan, who was the wife of Charles. In a 1920 letter from Santa Barbara, California, to Elizabeth's sister, Mary Whalley, Charles remembered this time and the blossoming of romance after Susan's death.

> It is nearly thirty years ago that Elizabeth was in my wife's mother's family in Cambridge, greatly

> beloved and remembered she was my wife's companion and we all adored her. Well, I always admired Elizabeth but I never thought of marrying her any more than a church steeple . . . But I could never forget her after she left us twenty years ago to come to California. I wrote her twice to tell her how I missed her, etc. but not a word could I ever get from her. My wife wrote to the same effect. but she dropped into a hole somewhere and after my wife died, I had a great time digging her out as a witness to my wife's will.

Charles moved to California, and he and his daughter hoped that Elizabeth would come live with them but Charles "thought it hopeless and that she would not give up her cafeteria in Porterville to live with us."

> I was very lonely separated from my son and daughter and my wife in her grave. I never thought of marrying Elizabeth for as much as I admired her, I did not love her. I was sitting by the seashore one day very sad as to my present condition and future prospects when all at once it seemed as if a voice spoke to me, 'Why not go to Porterville and try to get Miss Holmes to marry you?'

Charles goes on at length in his letter about his efforts to overcome Elizabeth's refusal, including mobilizing his son and daughter, until Elizabeth agrees ("I believe I can learn to love you though to love a man is something very different."). He closes by seeking to reassure Mary about his commitment to Elizabeth:

> You know what a treasure of a women your sister is and I want you to know that she will have a husband that understands and appreciates her as well as loves and adores her.[8]

Elizabeth was 59 when she married, her first marriage. As a married couple, Charles and Elizabeth made many visits to the Holmes farm, and several pictures survive, including the following (Figure 4.3).

*Figure 4.3 Charles and Elizabeth Stowe*

Charles and Elizabeth lived together for 13 years until Elizabeth's death in 1933. Elizabeth's final six years were ones of suffering, and she had a nurse with her throughout. Charles wrote a heart-rending poem, "My Elizabeth," in her memory. Among the archives of the Holmes family, is a beautifully illustrated "Young Folks Edition" of *Uncle Tom's Cabin*, a gift to Elizabeth's niece, Marion Holmes, from Charles.[9] The book cover (Figure 4.4) is shown below. Charles wrote an inscription to Marion (Figure 4.5; next page), in which he points out that his mother started writing the famous book on February 12, 1851.

*Figure 4.4 Cover, Uncle Tom's Cabin*

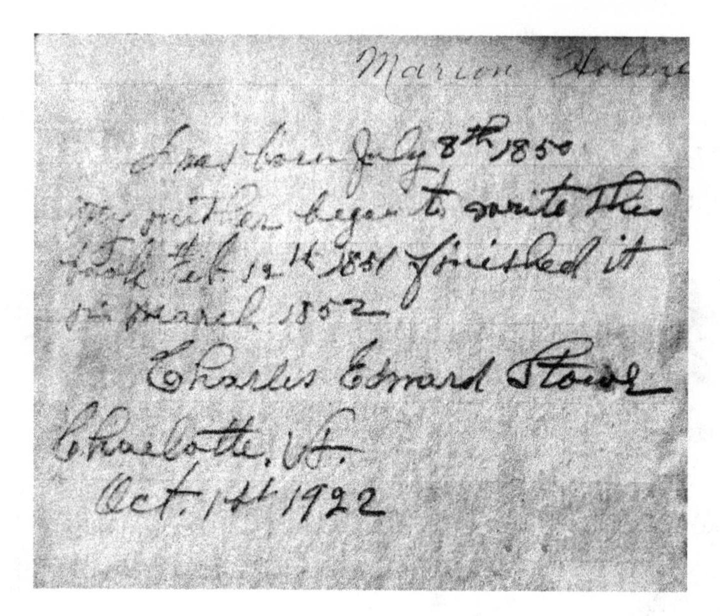
Marion Holme

I was born July 8th 1850
My mother began to write the
book Feb 12th 1851 finished it
in March 1852
Charles Edward Stowe
Charlotte, Vt.
Oct. 1st 1922

*Figure 4.5 Charles Stowe Inscription*

This trove of letters from the past gives us insight about earlier family members but we lack similar letters describing C.T. and Clara. The best insight comes from the robust story of the farm during the years they oversaw the farm and household.

## B. Schooling

Despite life focused on the business of the farm, the family valued books and education. Both in school and at home, a favorite pastime was memorizing and reciting poetry. But gaining a superior education was a challenge.

### *The Situation of Vermont Schools*

Concerned about the state of education in Vermont in the early 1900s, the Governor appointed a special "Commission to Investigate the Educational System and Conditions of Vermont." Its first and most important action was to hire the Carnegie Foundation for the Advancement of Teaching to assess the situation. The Foundation published a comprehensive report in 1914, which contained an excellent—and sobering—description of Vermont education of the period.

> It is to be remembered that Vermont, in larger measure than other states, has sent its young people away from home into the industrial occupations of other communities. Its population has remained practically stationary for nearly a half century. Its young people have gone to other New England states, to New York, or the west. With this migration there is no question that the system of education hitherto pursued has had something to do. In rural communities such as prevail in Vermont, the problem of the common school overshadows all others.

The study went on point out that "of the nearly 1,700 schoolhouses in the state, nearly 1,400 are one-room school buildings—nearly all of these rural."[10]

Of these 1,700 schools, only 979 were graded. Moreover, schools were small in enrollment: in 1912, 250 schools had fewer than 8 students, 552 schools had from 8–15 students.[11] Typically, the small schools employed one director/teacher for all grades and ages. Also, only after 1867 did public elementary schools become free. Paradoxically, despite the seemingly marginal education provided, Vermont ranked first nationally in the proportion of children enrolled in school and 5th in average attendance per child.[12]

Vermont families, such as the Holmeses, believed in education and made sure that their children made it to school.

Teaching was predominantly a female occupation: among 2,700 public school teachers in 1912–1913, 2,500 were women.[13]

> The lack in large sections of Vermont of occupations for women other than teaching has had a tendency to keep the teacher's salary very low, and this in turn has prevented rural schools in particular from

> obtaining teachers who have had a considerable amount of professional training. The typical rural school teacher is therefore a young woman of about twenty-three. She is a graduate of a four-year high school, but has no professional training.[14]
>
> More than half of the teachers employed in rural schools in 1912–1913 were new to the schools where they taught.[15] One hopeful sign was that the state had recently taken ownership of the two Vermont normal schools in Castleton and Johnson and increased its financial support of teacher training. Enrollments were 87 at Castleton and 56 at Johnson in 1911–1912.[16]

Thus, in the years that the Holmes children attended Charlotte schools (beginning in 1822), their education was characterized by one-room schoolhouses, small enrollments, inexperienced teachers, an ungraded curriculum, and frequent turnover of the school director-teacher. Yet, education was an priority of the family. Also, as seen below, the Holmes women aspired to continue their studies.

*Educating the Holmes Children*

One-room schoolhouses were prevalent in Charlotte during the 1800s and early 1900s. Despite the limitations of this educational structure (e.g., one teacher, typically inexperienced, for all levels and ages), there were strengths that reinforced the communal values of the time.

A longtime veteran of teaching in Vermont one-room schools, Wilma Farman, offered that her schools "had working groups small enough in number to have something of the old town-meeting quality" and her students "were saturated to the marrow of their bones by constant contact with the feelings of communal responsibility."[17]

> It worked very well to have older students help the younger students. It's more like a big family. Or someone whose work is done will ask if he can

> help Tommy with his multiplication table and they go off in a corner somewhere to do it. We'd have several things going on at once in the classroom. A one-room school is the same as an open classroom, but we didn't call it that. There's no better place than a one-room school for real teaching experience because you get to see everything in every grade.[18]

In Wilma's remembrance, she found that "Our children were more able to cope with questions. They'd discuss 'em with each other."[19]

Out of necessity, keeping the school functioning was a shared responsibility.

> Our pupils took turns doing the duties like cleaning the blackboards and sweeping the floor and cleaning the toilet and dusting and taking care of the bookcases. If we need any outside help, like the water froze, we'd call one of the fathers up. The fathers saved us a lot of money. They always put the paper around the foundation of the building to make it warmer in the winter. We tried to make school a social center because there wasn't much socializing back in the good old days. The kids came from farm homes and they brought their own hot lunches. Sometimes they brought 'em in a little jar. Mashed potatoes or maybe soup. I had a big old canner I put on top of the wood stove to heat the jars. That was our hot lunch program.[20]

From Wilma's oral history, one gets a sense of how the one-room schools were an integral part of rural life and its core values. The Holmes children were immersed in this experience. It was not a perfect educational arrangement, of course. As shown below in John Dewey's time as the sole teacher at Charlotte's Lake View Seminary, the relationship between teacher and school didn't always work out.

Jonathan attended his early years of school in Charlotte in a one-room school. Later, he traveled a few miles south to attend the Academy in Ferrisburgh, which had a larger, graded enrollment. The picture below (Figure 4.6), shows the Ferrisburgh school from around 1900.

*Figure 4.6 Ferrisburgh school*

Ferrisburgh was a strong Quaker community—stronger than Charlotte—so it made sense that Jonathan went to their school. The family of the famous author, Rowland Robinson, devout Quakers, had their farm just to the south of the school.

William and his wife, Mary Will, sent their two oldest children, Sherman and Edith, away to Castleton, for schooling with relatives. The children were probably in their early teens. Mary Will wrote that the children "have been in Castleton at school since last August. I have an uncle and wife and two aunts in the building with them. Uncle is principal."

The local schools were a fragile enterprise but a fixture in the lives of the children. Mary Will wrote in 1895 about the brick schoolhouse near the farm:

> John, Carleton and Hattie (her youngest children) are in school at the little brick school on the corner which the trustees are trying to keep together through this school year as the new law goes into effect in the fall, and they are talking of throwing up some of the district schools and establish a grade school at the corner.

The picture below (Figure 4.7) shows the nondescript school building in the 1880s, with Robert Holmes and his cousin, Edith, sitting in front. With the Holmes family committed to this small school, it is probable that the bricks came from the nearby Holmes brickyard.

*Figure 4.7 Brick school, Charlotte*

About 30 years later in 1922, Robert's two children, Marion and John, attended the same school, which became known as the Old Brick School. In the picture below taken on a winter day (Figure 4.8), John is seated on the steps, and Marion is standing third from left.

*Figure 4.8 Children at brick school*

By this time, it was designated as a graded school. Marion's monthly report card for third grade in 1919–1910 has this reminder to "parents and guardians" printed at the top:

> The records of every school show that greatest cause of poor work in school is irregular attendance. A written note to explain each absence of your child will be appreciated by the teacher and will assist her in her efforts for the welfare of your child.

The obvious assumption is that teachers would be female. Marion's teacher, Mildred Tupper, indicated that Marion had missed 15 days of school in the 8th and final month of the school year. Whether Marion had pressing duties at the farm or had an illness is not known. For nine rated areas (including conduct), Marion had nine A's and a B in arithmetic and was promoted to fourth grade. [21]

There was another Charlotte school that enrolled several Holmes children over the years. From 1840–1880, the Methodist Episcopal Society sponsored a school, but it was destroyed by a fire. Local citizens then raised money to build a new high school called the Lake View Seminary, which opened in 1881 (Figure 4.9). An interesting historical footnote is that John Dewey was hired to oversee the stately new school for the winter term, 1881–1882.

*Figure 4.9 Lakeview Seminary*

A brilliant young man of frail stature, Dewey grew up in Burlington and attended the University of Vermont where he studied under eminent scholars, such as H.A.P. Torrey. He harbored ambitions as a scholar and went on to graduate school. He received a PhD and earned world-wide fame as a philosopher, teacher, and author. George Dykuizen, a Dewey biographer, wrote the following about Dewey's experience with the farm boys of Charlotte:

> The school had between thirty and thirty-five students, ranging in age from 13 to 20, mostly from farm families. Many were ill-prepared for the work they undertook and belonged more properly on lower levels of study, which led to much frustration for both pupils and teacher, creating a general opinion among the townspeople that Dewey's teaching was not "average good."

While most of the children were well-behaved, the older boys were mischievous and unruly, and played all manners of pranks on each other and their teacher. Dewey's attempts to control the situation were only partly successful.[22]

We don't know whether among the "unruly" farm boys was a Holmes. At any rate, Dewey left after the winter term and enrolled the next fall at Johns Hopkins University to pursue graduate study. Knowing his ambitions as a philosopher and his less than successful foray as a teacher and school head, it was a posting that was not to last anyway.

*Women's Roles and Teaching as an Outlet*

Strong-willed Hannah Holmes and the other Holmes women had active minds, superb organizational skills, and the ability to solve the practical problems of the day. They were readers and innate learners. They were capable of many roles and occupations.

Their futures, however, were constrained by the cultural expectations of the time. One writer characterized the 1800s as follows: "A time influenced heavily by the rise of the domestic ideology or code that "prescribed expected behavior by white native-born women—they were ultimately to be full-time mothers."

The choices for young women were guided by both the reality and the perception of their growing importance in the domestic arena. The decision to take employment outside the home, and what employment to take, was increasingly judged in terms of how it would prepare women for their roles of wife and mother.[23]

Another observer noted the most common outlet beyond the home:

> The same moral code that denied the legitimacy of wage work provided justification for opening up educational institutions on the ground that educated women made better mothers, and it offered a rationale for women to become teachers.[24]

Teaching, however, was rarely a lifetime commitment. It was common for a girl who started teaching as a teen to stop once she married. Thomas Dublin, in his study of New Hampshire teachers in the late nineteenth century, found that 60 percent of women teachers had ended their career by age 24.[25]

Several Holmes women sought studies beyond the limited curriculum of the local schools, studies that took them away from life on the farm. As noted above, Jonathan's daughter, Elizabeth, pursued a music career and married into the Stowes, one of America's most renowned literary families.

The daughters of C.T. and Clara left the farm to become educators. Maude moved to Texas and became a school director. Hannah Elizabeth graduated from Burlington High School and was a Phi Beta Kappa graduate of the University of Vermont in 1906. After graduation she became "preceptress" at Montpelier Seminary. She was a math teacher, a critical source of income for her family when her husband, Jacob, was stationed in Europe during World War I and then died at a young age.

Rena, who married C.T.'s son, Robert, followed the more typical pattern of those entering teaching. She received her teaching degree in 1900 from the State Normal School in Johnson. Although Rena eventually married Robert and came to the Holmes farm, her teaching career was a stark alternative to spending her entire life on a Vermont farm.

As an 11-year-old living in Bristol, 20 miles southeast of Charlotte, Rena wrote a letter to her cousin, Ina, on July 25, 1893. It was often the case that children were sent from relative to relative during hard financial times. Recognizing the expected writing lapses of a young girl, Rena expressed her unhappiness about being in Bristol:

> We are well, so are Uncle Frank people except him he has the teeth ache. You asked me about how I liked to live down here, well I don't like it near as well as I did up there in Mr. Hosford's house. But I think it is because there are no young people down here like there were up there.
>
> Yes, I remember how Charlie (Rena's brother) shut us in the ice house. I wished we lived there now so he could now. I remember coming home on the snow banks and how we stayed there a long while after school was out nights to slide. Yes, we have our haying done. We finished this morning and it is lucky too because it is raining hard. My teachers name was Alice summer. She was only 17.

Fifteen months later, on October 30, 1994, Rena wrote Ina again and alluded to some unhappiness about school:

> How is your liver these days? Mine isn't very well. I am not very sorry that my school is most out. It isn't because I don't like it but I am tired of going to school. How many studies do you take Ina. I suppose you take a good lot.
>
> Wednesday day eve, oh how it rains. Mama has gone to West Charlotte Vt to do that Sunday School convention today . . . Mama has just come in. The scholars who study Shakespeare are going to give a reading Friday evening. We expect it to be nice. Supper is ready so must go. Well, I must close and work on arith.[26]

Four years later in 1898 at age 16, Rena started her studies at the State Normal School in Johnson in northern Vermont. She traveled by train each week to Johnson and boarded there during the week. As Rena told me many years later, she would pack food for the week and carry it to Johnson on the train and make it last for five days.

Johnson Normal School included a two-story frame building for instruction and a dormitory a half mile away where Rena boarded during the week. Near the dormitory, the village public school served as the practice school for the students in training.

Rena, now 18 years old, sent a formal invitation (Figure 4.10) to the family for the graduation ceremony on June 14, 1900.

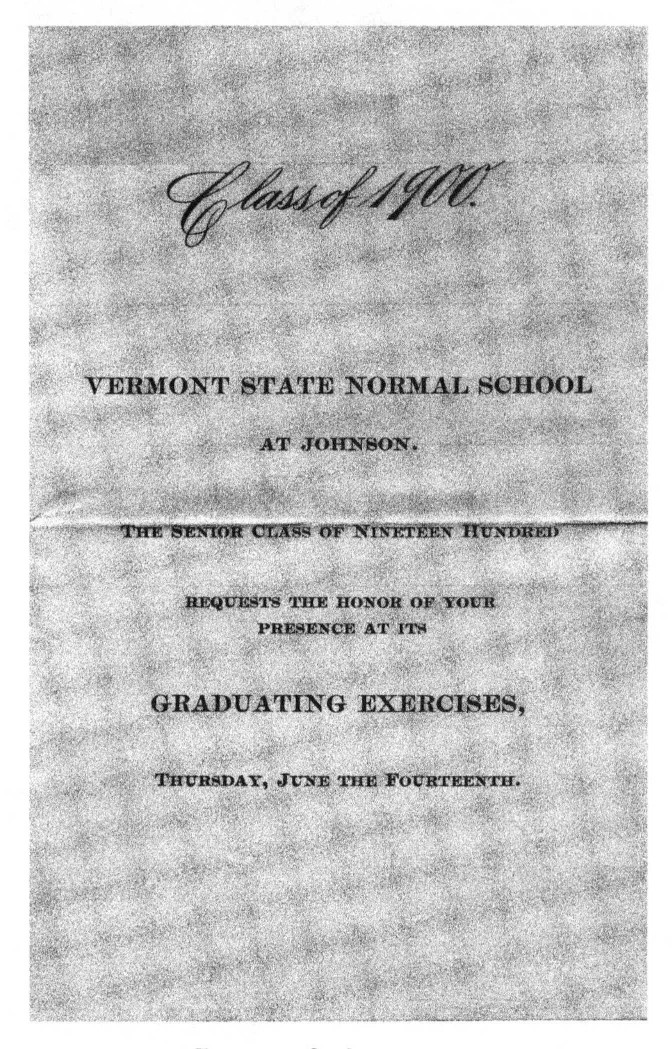

Class of 1900.

VERMONT STATE NORMAL SCHOOL

AT JOHNSON.

THE SENIOR CLASS OF NINETEEN HUNDRED

REQUESTS THE HONOR OF YOUR
PRESENCE AT ITS

GRADUATING EXERCISES,

THURSDAY, JUNE THE FOURTEENTH.

*Figure 4.10 Graduation program*

Rena's graduation picture (Figure 4.11) shows a serious young woman about to embark on a teaching career.

*Figure 4.11 Rena's graduation photo*

Rena taught school for 10 years until 1910, when she returned to Charlotte and married Robert Holmes on June 7, 1911. She was almost 29 when she wed, close to what was considered spinsterhood at the time. We don't know anything about the courtship—perhaps Rena and Robert maintained a correspondence while she was away and saw each other on holidays back in Charlotte. Perhaps, too, Rena's exit from teaching coincided with an inner sense—and perhaps family pressure—that she should get married while she was of marriageable age and raise a family.

We know, however, that Rena had a stellar 10-year career. She started out in the Charlotte graded school for four years, taught in Hanover, New Hampshire, for four years, then spent two years in Indianapolis, Indiana. In her final year in the Charlotte school, she received an award from The Burlington Free Press that included a trip to the World's Fair in St. Louis.

The Free Press Manager wrote Rena on June 24, 1904, that, after reconsideration, she should get the trip:

> As you will see by the Free Press in the morning, Miss Delano had a large lead in your class, but the vote which you had was so large that we would be pleased to furnish you with a ticket to St, Louis, just as if you had won.[27]

In hindsight, we don't know the machinations that led to the change of heart by the Free Press, but family members remembered the honor many years later.

At different points in her career when she was seeking new employment, administrators wrote glowing recommendations. In 1903 the former head of the teacher training school in Johnson, who followed her career, wrote on her behalf at a time when she was looking to leave the Charlotte school:

> All children take to her at once and she wins them through her own love for them. She is a very attractive young woman; some would call her charming in personality and all would agree that she was cut out for a teacher who would win her way in the hearts of all acquaintances.[28]

When she decided to leave the school in Hanover, she had letters from a member of the school board and from a professor in the department of Philosophy at Dartmouth College. Charles Adams, the board member, wrote:

> Miss Johns has the natural gifts of a successful primary teacher; she is refined and quiet in a manner, sympathetic in her handling of little children, a good singer, and resourceful in method. Her success here is certainly such as to justify her in seeking a position of larger responsibility and more generous income.[29]

H.H. Horne, the Dartmouth professor and probably also a board member, spoke about her "winning the love" of her students and her other attributes:

> The lessons she teaches carry a training in character as well as conveying interesting information. She is a Christian lady of great attractiveness to her pupils, their parents and her fellow teachers. Hanover is losing her because it cannot offer enough to keep her here.[30]

Rena was clear to her employers that money was a concern and a motivation for looking for a better situation. It speaks to the value of the dollar in 1908 that she accepted an offer in Indianapolis for an annual salary of $750 "payable in ten monthly payments."[31] Yet it was a nice increase over the pay in Hanover. Rena's practical, bottom-line mentality came through later when she played a significant role in family matters, particularly in how to respond to the financial downturn of the farm.

Rena's affection for her students—and instinct for whimsy—comes through in this picture of her Hanover students performing a dance (Figure 4.12), with Rena peering from the front door of the school.

*Figure 4.12 Rena's students*

## C. Fresh Air, Fun, and Family

A reality of living in Vermont, especially on the windblown lake, was dealing with the harsh winters. Before the rise of global warming, Lake Champlain froze over—all 120 miles—almost every winter and was a basic avenue of transportation. The Holmeses rode a sled to Burlington to conduct business and do shopping. Yet, even then, the old adage—"if you don't like the weather, just wait a minute"—held true. There were beautiful sunny days, many above freezing, and these days brought people outside and had a therapeutic effect on morale.

Yet, the winters were long (October–November to March–April) and had long spells of harsh weather. Author and chronicler of life in Vermont, Murray Hoyt, wrote about living along the lake on Potash Bay (25 miles south of Charlotte) in the first half of the twentieth century and conveyed these memories of winter:

> The world of winter on Lake Champlain is just as beautiful as the summer world, but in an entirely different way. It's a white world trimmed in black. The vast expanse of snow-covered ice is blindingly white in bright sunlight, and beyond the lake the snow-covered mountains have splotches and lines

> of black, which are the dark evergreens . . . Always, though, even in the brightest sunshine, the scene looks incredibly cold.[32]

Hannah noted in February 1881:

> We had a great depth (of snow) for a short time and then the weather was so cold there was no comfort in going anywhere, then two weeks ago the south wind began to blow and kept up until the ground was nearly bare.

But the activities of the farm went on. Roads needed to be plowed. This picture (Figure 4.13) shows C.T.'s son, Robert Holmes, smoothing out and packing the snow.

*Figure 4.13 Plowing snow*

For all the work that needed to be done, however, an act of survival in winter was getting away from the wood stoves and stale air of the house, combatting cabin fever, and finding ways to enjoy and take advantage of the winter months. In the following picture (Figure 4.14) Robert prepares his children, Marion and John, for a ride on the sled.

*Figure 4.14 Robert and children on sled*

The picture below (Figure 4.15) shows Marion, at about age 4, with snowman, sled, and baby brother, taken from the Robert Holmes residence looking northeast toward C.T.'s house.

*Figure 4.15 Snowman*

Getting outdoors for a sleigh ride worked wonders for morale. Clara, bundled-up with a fur coat, heads out, drawn by two sturdy horses (Figure 4.16; next page).

*Figure 4.16 Horses and sleigh*

Iced-over Lake Champlain brought skating and sledding. Robert got Marion and John on skates at an early age (Figure 4.17).

*Figure 4.17 Skating on Lake Champlain*

Getting onto the ice was a unique sensation. As Murray Hoyt commented,

> It seems astonishing to a rink-skater to start off in one direction and continue in that direction indefinitely. But this isn't the only difference. The

> biggest difference is in the effect that wind has on you. When you skate into even a light wind a huge, invisible hand seems to hold you back . . . But when you skate *with* even with a light wind your strokes lengthen and your speed increases unbelievably. You seem to fly along.[33]

Also, in 1900 at the age of 5, Robert was introduced to skiing. His father, C.T., made him a pair of skis out of barrel staves: "He made me a single jumper that was a thrill for many years." Undoubtedly, C.T. and Robert introduced Marion and John to barrel staves and the idea of skiing. As described later, John became a top competitive skier after the move to Middlebury and was captain of the Middlebury College ski team in 1936.

When mud season ebbed and summer arrived, the family had many more —and easier—options for getting outdoors. Horse-drawn carriages—driven by the men and women—transported the family to the store, nearby relatives, church, and picnics. In addition to training and racing horses, riding a horse was commonplace for everyone. For children, learning to ride a horse was as natural as learning to walk. The children often got started with a pony and small carriage. Below (Figure 4.18), John Holmes, his cousin Austin Ross, and a neighbor girl are trying to get their pony, Pet, to move.

*Figure 4.18 Kids and pony*

The family adored their children. Two girls, Marion Holmes and Katherine Ross, were posed for this portrait (Figure 4.19) with one of the barns in the background.

*Figure 4.19 Two girls*

There were games too, and in the picture below (Figure 4.20) C.T. is taking a croquet swing.

*Figure 4.20 Croquet*

Lake Champlain brought boating, swimming, and picnics. Following Sunday church, C.T., seated at right, entertained in-laws at the lake shore. The picture below (Figure 4.21) shows C.T. with members of the Johns family. The Johns connection was made when C.T.'s son, Robert, married Rena Johns.

*Figure 4.21 Sunday beach outing*

It does not appear that hunting and fishing were priorities of the family, either for recreation or for bringing food to the dinner table. We know, however, that guns were a part of life on the farm, and people were comfortable with them. When an animal had to be put down, a rifle would be used. Rena Holmes, who grew up in Charlotte and lived at the farm after her marriage, kept two small pistols. She showed them to me and told me that she used a pistol for shooting squirrels when she was a girl.

Once, on a visit to my grandparents when I was 11 or 12, I wielded my new Red Ryder air rifle to shoot a chipmunk behind the Middlebury house and received a sharp reprimand from Rena.

The family's ease with guns was evident in later years. Rena's husband, Robert, did occasional deer hunting after they moved to Middlebury and had a deer rifle for this purpose, but this was not a passion for him. When I visited Middlebury as a boy, he took me to the town dump where we shot rats. He taught me how to use a 22-

long rifle for that purpose. I remember too how natural it was at the family camp at Lake Dunmore to use a rifle to shoot a porcupine out of a tree. They were worried that the dogs would be hit with quills, which had happened more than once.

Later, when their son, John, went into the FBI, wielding a weapon came naturally. As it turns out, I am the repository of the family "arsenal." I have Rena's two pistols (Figure 4.22), a third small pistol, Robert's deer rifle, two 22-long rifles, and John's .38 caliber FBI pistol.

*Figure 4.22 Rena's two pistols*

The 1800s were a time of steamboat traffic on Lake Champlain, and the Holmeses had a balcony seat to watch the action on the lake. As Ralph Nading Hill wrote: "Travelers between Montreal and New York preferred water transport to the cord or dirt roads over which the stages bumped. Shippers found the steamers swifter and filled their holds with quantities of merchandise."[34]

The Champlain Transportation Company, headquartered in Burlington, was the leader in the vigorous competition to command the shipping business on the lake. The company launched the 120-foot *Vermonter* in 1809, which traveled from Montreal to Whitehall, New York, and other builders jumped into the fray.

The Shelburne shipyard, which was founded in 1820, built an even grander boat in 1837, the Burlington. The shipyard, just 10 miles north of the Holmes farm in Charlotte, built a total of 12 sidewheelers, the last being the *Ticonderoga*. Dan Sabick, an expert on the history of steamboats and director of archaeology at the Lake Champlain Maritime Museum, points out, "Lake Champlain was instrumental in the development of steam-powered technology and the Shelburne Shipyard was a large part of that, and was really kind of a hot bed of experimentation for steam technology."[35] Sabick noted also that numerous steamboats were scuttled in Shelburne Bay and expressed concern about their preservation.

Steamboats dominated the shipping business until the railroads emerged later in the 1800s as the primary transporter of goods. A total of 29 steamboats traveled the lake over more than a century.[36]

Although the Holmes family shipped produce on the lake, they used other boats for that purpose. Moreover, in time, the steamboats became more of a passenger vehicle, and members of the family took holiday excursions on the boats. Here is a picture (Figure 4.23) of the steamship *Vermonter* taken by a family member from the shoreline.

*Figure 4.23 Vermont steamship*

The 220-foot *Ticonderoga*, built in 1906, sailed the lake until it was moved several miles on a make-shift rail from the shoreline to the Shelburne Museum in 1955. The idea of saving the boat was the inspiration of the Webb family who founded the museum. As the museum asserts, the move was "a remarkable engineering effort that stands as one of the great feats of maritime preservation."[37] In the summer of 1953, my grandparents took me, age 11, on one of the last rides of the *Ticonderoga*. The circular for rides that summer is shown below (Figure 4.24).[38]

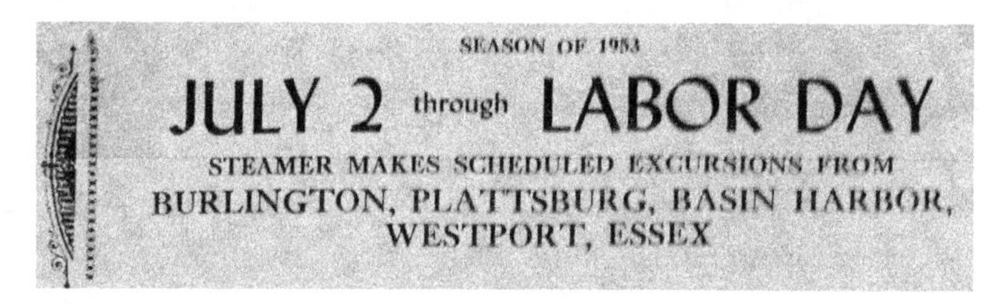


*Figure 4.24* Ticonderoga *trips broadsheet*

Below is the steamer's final resting place at the Shelburne Museum (Figure 4.25).

*Figure 4.25* Ticonderoga *at Shelburne Museum, Photo by Storylanding*

Gathering groups of people on special occasions came naturally to the family. The family hosted Vermont's Congressman Porter Dale for a campaign speech from the porch of the homestead (Figure 4.26).

*Figure 4.26 Porter Dale speaking*

A few years later, Dale earned a footnote in American history. In 1923, while campaigning for the U.S. Senate, Dale heard of the death of President Warren G. Harding and traveled immediately to Plymouth to ensure that Vice President Calvin Coolidge knew about Harding's death. Several accounts indicate that Dale urged that Coolidge be sworn in immediately to ensure continuity in the presidency. A day later, on Aug. 3, Dale witnessed Coolidge receiving the oath of office.[39]

It says something about the stature of the Holmes farm and family that Vermont's sole congressman chose to campaign at the farm. By this time, of course, C.T. was widely known statewide for the apple orchard. In the next chapter, we will see that in 1911 C.T. and Clara hosted the annual summer meeting of the Vermont State Horticultural Society. The occasion attracted 350 guests.[40]

Lacking the tools of modern communication or even reliable telephoning, the family communicated by letters and, on numerous occasions each year, at family gatherings. These occasions included Sunday church, Christmas and Easter holidays, the 4th of July, race days at the track, marriages, and funerals. Because most members of the extended family were in geographic proximity (East and West Charlotte, Ferrisburgh, and Monkton), gathering the family was feasible. Even after William and his large family moved to Proctor, about 40 miles to the south, people made the trek in both directions.

The move off the farm in 1898 by William and his family was a major blow to the family. Even before Jonathan died in 1894, Jonathan had been unable to play an active role in running the farm due to his disabling condition. Thus, for many years, C.T. and William were the lead partners in the enterprise. The story and pictures of the apple orchard in the next chapter show vividly the important role of William.

Mary Will, Williams's wife, captured the financial challenges in a letter of January 30, 1899, including the harsh economic realities of farm life: "We had economized, done without, and worn out most of our clothes trying to stay on the farm, hoping to have a crop of apples. Everything was put into the orchard, but as there were no apples this year we could not stand it any longer. Charlie picked less than 50 barrels of No. 22 when we left."

So, William, Mary Will, and their seven children moved to Proctor, where William took a job with the flourishing Proctor Marble Company. It was a tough time for William, but it could have been worse—he might have been sent to Mexico.

> We came to Proctor the first of October. It is so very different than it ever has been. Will is here at meal time and through the evening, when he comes from his work at twenty minutes of five he is very tired, usually goes on to bed and lies until called to supper. After supper may take his violin and perhaps read for an hour then goes to bed. He used to do so much for the older children but is too tired to even read to them or tell stories. Perhaps he may become accustomed to this after a time and not be so tired, and perhaps his work will be lighter for they cannot be putting heavy shafting all the time.

> It is comfort to have him come into his meals three times a day rather than have him down in Mexico, as he expected to go when we came here.

Mary Will, who was proud of their new house back at the farm, realized quickly that life in a factory town and life at the farm were profoundly different:

> It seems too strange to me to be paying for a stove by $5 a month, for on the farm, usually, something is sold and such things paid for in a lump. We left pretty much everything behind us, brought my best furniture and all we could use in this little house, packed the rest in my spare chamber and locked it. The house belongs to a private party. We hope in the spring to have of the company's houses with modern improvements.
>
> These poor men, how they are abused. I wonder if they lose all their individuality. You and I have our own private opinion I am sure.
>
> Perhaps you can imagine how I feel after having lived all my life in one town, growing up with the people, to go out among people and seeing all strange faces.

Another reminder of how things had changed was Mary Will's finding that "Charlie (C.T.) has rented our house and the man works for him."[41]

William and his family were gone but they were not forgotten. His niece, Marion Holmes, daughter of Robert, talked affectionately about "Uncle Will" for the rest of her life and remembered the family gatherings that occurred after the move. One such occasion was Christmas Day in 1912 with William and his family back at the farm. William, in his full beard, is at the far upper right next to his brother, C.T. (Figure 4.27).

*Figure 4.27 Christmas Day, 1912*

Thirteen years later, in 1925, the family celebrated William and Mary Will's 50th wedding anniversary. William and Mary Will hosted the event in Proctor for 60 relatives. It was the last reunion of those who had lived at the farm. Pictured below (Figure 4.28) is the gathering of the family that had remarkable continuity in the same place for 101 years, but was now scattering. Four years later, William died; five years later C.T. died.

*Figure 4.28 50th wedding anniversary*

Life on the Holmes farm included many people (family and hired hands), a complex business undertaking, reacting to regional and national economic forces, the vagaries of weather, the power of family bonds, and the joys of building something important. It was not for the faint of heart. These were hardworking, entrepreneurial, resilient, physically strong, god-fearing, loving, supportive people.

From all indications, they enjoyed living in this time and place. Day by day, there is something infinitely tangible about farm work: you do it, you see it, you touch it, you share it. And you go to bed at night—dog-tired or not—with an authentic sense of accomplishment. The story of the apple orchard exemplifies these characteristics.

Chapter Five

# Ascendance of C.T. and the Orchard

*"On Mr. Holmes' farm it is the rule to attend to the requirements of each tree individually, to manure the ground, to prune the tree and spray it as if that were the only tree on the farm."*

Jonathan, age 42 at the time, planted 100 acres of apple trees around 1862. This was the start of a 50-year undertaking that culminated in regional and national prominence by 1910. Just a boy when the plantings occurred, his son C.T. worked year after year beside his father until Jonathan's debilitating illness took effect. Knowing it takes 40 to 50 years for an apple tree or an orchard to mature, it can be said that C.T. grew up with the trees. The picture on the facing page (Figure 5.1) shows the handsome young man at about age 22, circa 1880.

*Figure 5.1 C.T. as a young man*

## A. Professionalization of Apple Growing in Vermont

The Holmes orchard emerged during a time—around 1900—that Vermont began to recognize the economic potential of apple orchards. The number of commercial orchards statewide was growing, and the science of apple culture was taking hold. As will be seen, the Holmes family played a role in the professionalization of the Vermont apple industry.

Prior to this time, however, it was a different story. M.B. Cummings, the state's leading expert on apple production and a promoter of the Holmes orchard, surmised that "The early orchards were probably spontaneous in origin, coming from wild seedlings and not from grated trees."[1]

### *Hard Cider and Temperance*

Cummings also noted that, beginning in the early 1800s, large quantities of apples were grown across the state to produce a singular apple product, cider. It was said that people drank cider as freely as water. It did not take much Yankee ingenuity to develop a version known as hard cider, with a further offshoot called applejack. One writer observed the following:

> Applejack was traditionally produced from the hard cider that was the drink for most Americans in the 18th century. Naturally fermented and low in alcohol, hard cider was safer than well water, cheaper than beer, and easy to make at home.[2]

Many historians believe applejack was invented by American colonists. It was produced by a low-tech method called "jacking." Jacked spirits are distilled not by the usual method of boiling but by freezing, and any household with a supply of hard cider and cold weather could make applejack.

Along with the disturbances caused by drunkenness, health risks were evident. Home brew "frequently contained hazardous compounds, like methanol and acetone, that can cause blindness, renal failure, and other permanent damage." The health risks made apple orchards a target of the temperance movement, starting in the 19th century. In 1884, as the movement gathered strength, the New York Times published an editorial titled "A Wicked Beverage," condemning not only liquor in general, but applejack in particular. "The name has a homely, innocent appearance," it said, "but in reality, applejack is a particularly powerful and evil spirit." By the time Prohibition ended in 1933, millions of acres of orchards had been razed and were never replanted. Cheap and plentiful grain made whiskey easier to produce.[3]

To the chagrin of Vermonters with a temperance orientation, cider in its hard form was hugely popular. In 1810 there were 125 distilleries in Vermont, with cider brandy a primary product.

According to Cummings, the hot market for apple-based alcohol didn't last, and the number of distilleries fell to just two by 1840.

> The sterilization of sweet cider by boiling was unpracticed in the early days, and there was a considerable period of intemperance from 1810–1840. After a time, however, the spell was broken, and soon the farmers began to put their best apples to better use by feeding them to their cattle, their hogs, and horses.[4]

Temperance was a topic on the minds of Vermonters from the hard cider era into the 20th century. A state publication in 1893 bragged, "Vermont was one of the first states to enact laws restricting the sale of intoxicating liquors, and makes temperance one of the subjects to be taught in common schools."[5]

As for the Holmes farm and orchard, their consumption is difficult to discern. We don't know the extent to which, if at all, the Holmeses made hard cider or applejack, although the conditions (lots of apples, ease of boiling down cider, freezing temperatures for "jacking" the cider) were ideal.

A clue, perhaps, is the antagonism of my grandparents, Robert and Rena, toward alcohol consumption. They lived at the farm and may have experienced—been repulsed by—drunken behavior by members of the family or the hired hands. For example, an incident described below indicates that during picking time in 1910, with as many as 100 hired hands at the farm, excessive alcohol consumption occurred among the transients.

Robert and Rena internalized the temperance idea and, according to Robert, "never touched a drop." Their son, John, who was born on the Holmes farm in 1915, thought differently. He consumed alcohol in college in the 1930s, both before and after prohibition ended. In fact, he and a fellow student, a future United States Senator from Vermont, regaled me with a story about skiing and partying in the Mad River Valley, then returning to the Middlebury campus over the mountains in a snowstorm. They made it home safely thanks to the future senator, who was spread-eagled on the top of the car and yelled directions as they maneuvered their way home.

## *Evolution of Vermont Apple Industry*

Eventually, the apple market evolved from being for cider-lovers and feeding farm animals to becoming a staple item for families everywhere. Looking at this span of history, there were three phases in the development of the Vermont apple industry:

1. The apple cider period, which covered the time from the early 1800s and the Civil War to about 1875.
2. The family orchard period, which went from 1875 to about 1910.
3. The commercial period, which covered the years after 1910.[6]

The Holmes orchard spanned all three periods, from Jonathan's setting the orchard during the "cider era" to C.T.'s commercial successes in the early 1900s.

There were three main apple-producing regions in Vermont: The Connecticut Valley, the Champlain Valley, and the so-called Hill Town districts. A 1911 report disseminated by the Vermont Department of Agriculture, *Vermont: An Apple Growing State*, highlighted the Champlain Valley:

> Apple culture in this valley had been so well developed, comparatively speaking, that a general reputation is already established. Here commercial orcharding really began in Vermont. There are three large apple-producing counties in the Champlain Valley. They are: Addison, Grand Isle, and Chittenden. The largest orchard in the state is in the last-named county, and is owned by C.T. Holmes of Charlotte, whose immense plantation embraces over one hundred acres.[7]

The report went on to recommend six fruit-growers for expertise on apple growing, including C.T. Holmes, and featured a picture of the Holmes orchard. The bulk of the report contained advice on critical factors in developing an orchard: planting, pruning, tillage, cover crops, and pests.

During the early 1900s, there was a concerted effort to highlight the commercial potential of Vermont apples, and the natural growing conditions were perceived to be an important advantage. The 1911 report described several advantages:

> Vermont is a natural apple state and lies well within the apple belt of the United States. We have a climate which is congenial to the apple, abundant rain in most seasons, a warm summer, and a winter not too severe.
>
> However, the winters are cold enough to assist in repressing insects, pests and fungus diseases, which are less in evidence than elsewhere in this country. The soil is well adapted to the apple tree, generally well drained, for the most easily tilled, and seldom too scant in fertility. Granite and limestone soils prevail, and these are conceded to be congenial

> to apple roots. Grafted as well as native trees live a long time in our state.[8]

At about the same time, Cummings noted:

> Vermont is rapidly becoming a state in which apple growing is developing on a commercial scale. Vermont's natural advantages as an apple state are numerous, and include the following: superior flavor, excellent keeping qualities, favorable climate, good reputation, proximity to markets, good apple soil, and comparative cheapness of orchard land.[9]

An orchardist from Vergennes, E. N. Loomis, was even more fulsome in his assessment ("The Champlain Valley is the greatest apple producing section in the world.") but rued that Vermont was slow to respond:

> Strange to say the Vermonters alone but little value this fact and are indifferent to the possibilities of wealth presented by this great opportunity. Apple merchants know this but Vermonters have little faith and are slow to act.[10]

Through their annual reports and meetings, the Department of Agriculture and the Vermont State Horticultural Society reinforced the message about the potential of apple growing. In addition, they became important repositories and conveyers of scientific knowledge about apples.

To complete the picture, long-term challenges of commercial apple-growing in Vermont were also identified by the State experts, including: the up-front costs of building an orchard, the many years until there is an income flow, competition from other eastern states, the inevitable vagaries of weather, the increasing cost of land for creating new orchards, the expectations of urban consumers that an apple be perfect in appearance.

Writing in 1911 on the latter issue, a University of Vermont professor, B. F. Lutman, pointed out an ironic twist: building the consumer desire for apples brought a heightened expectation for the quality of the product and higher costs of production:

> Twenty-five years ago, the town folk were glad to get the farmer's apples at the low price then asked for them. They accepted them as they bought them and were glad to get them. They were not perfect usually; they were scabby and probably often wormy . . . The city dwellers were willing then to accept the product. Soon better methods of caring for the trees and fruits came into being—better methods of pruning, of spraying (before unknown), of picking the fruit and of packing and marketing it. Perfect fruit began to appear on the markets and the imperfect fruit could not compete with it, but the city dwellers wanted the new perfect fruit at the old cheap price. Pruning, care, fertilizing, spraying, and proper packing take time and cost money.[11]

Lutman's answer to the financial squeeze on the apple producer—higher costs of production and the need to sell apples at a higher price to the producer—was to ensure a top-quality product through "hygienic measures" and "spraying measures" that would combat the scab condition, the codling moth, the railroad worm, and scale insects. His presentation included a picture of C.T. Holmes's demonstration of a power sprayer at the 1911 summer meeting of the State Horticultural Society. Lutman offered a list of recommendations that suggest the dedication and expertise required to maintain and improve an orchard:

*Hygienic measures*

1. Gather up fallen fruit and feed it to stock or turn stock into the orchard and let them gather it up.
2. Rake up dead leaves and burn them.

3. Prune out all canker and apply a strong disinfecting spray to the limbs and trunk of the tree.
4. Prune close and paint the scar.

*Spraying measures*

1. Spray just before the blossoms open with a Bordeaux mixture or a lime-Sulphur mixture for scab.
2. Spray after about three-fourths of the petals have fallen with the same Bordeaux or lime-Sulphur mixture to which 2–3 pounds of arsenate of lead has been added; spray this time under at least 125 lbs. pressure and try to drive the spray into the flowers. This is for coddling moth and scab.
3. Spray in about two week or three weeks with Bordeaux or lime-Sulphur mixture for scab.
4. Spray in about three or four weeks more with the same mixture to which arsenate of lead had been added in order to kill the second brood of codling moth.[12]

If nothing else, this list indicates that growing apples for commercial purposes was a complex, serious business.

The development of the Holmes orchard coincided with the promotion of the Vermont apple industry and a concerted effort state-wide to advance the science of apple growing. Drawing on the words of C.T., the next section describes how the Holmes orchard came into being.

## B. Building an Orchard

In reconstructing the distant past, it is nearly impossible to know what people said and how they said it. A gift from more than 100 years ago—the early 1900s—comes from the Vermont State Horticultural Society. For an historian, a truly valuable contribution of the Society was that presentations were transcribed and published in the society's proceedings, which may be found in bound volumes in the Special Collections section of the University of Vermont library. From these sources, we have extensive transcribed remarks by C.T. Holmes and others about the orchard. In particular, we know 110 years later what C.T. said about the farm. We know how he thought and how he expressed himself.

Here is C.T. speaking about the orchard in 1914:

> I think the proper thing would be to go back about 52 years now and tell you why and how the orchard came to be set out. Father (Jonathan) had an old Greening tree and one fall he picked off 10 barrels and sold them for $5 a barrel. That was the start. The next spring, he began setting out trees and set 3700 Rode Island Greening, and the balance Baldwin and Spies to make the 5,000. That was practically 50 trees to the acre.
>
> Father knew nothing about the care of trees. It is evident that the soil condition and location are adapted to fruit trees, because if they were not I don't think there would be a dozen trees standing there today to tell the tale, but he set them out and took care of them as best he could. They were all in sod. Once in a while he plowed up and sowed a wheat crop and got what he could. We always sprayed the trees well, and that was I think what preserved the life of many of them, but it was a struggle to keep the (insects) away.

C.T. had something is say, too, about how his father planted the trees.

> Setting them at 50 trees to the acre made them 25 feet apart each way. He gave little thought that in 30 years or 35 that those trees would be practically

> interlocking, but it proved to be so, and my brother conceived the idea of cutting out half the orchard. Father was opposed to cutting out part of the trees, but we prevailed on him and we went to cutting out trees that would measure from a foot to eighteen inches in diameter; we persevered and cut every other row. That left rows 50 feet apart and the trees 25 feet in the row. That was an improvement for we could get in the orchard as it should be.[13]

Although Jonathan was disabled for many years and unable to oversee operations as he did in the early years, removing half of the trees must have been a hard move for him. Below (Figure 5.2) is a view of the orchard, looking from the south, before the cutting.

*Figure 5.2 Orchard view*

Gradually, however, his sons, especially C.T., asserted themselves. Below (Figure 5.3) is the newly-spaced orchard in full bloom.

*Figure 5.3 Orchard in bloom*

An article in the January 1911 edition of *The Garden Magazine* recapitulated the early history of the orchard:

> Almost like a fable runs the story of Mr. Holmes' orchard. Although the old farm had been in the family so many years, not until 1870 was much attention paid to the apple. At that time 100 Spys, Greenings and Baldwins were bought at a cost of $10. Even in the beginning great care was exercised to dig large holes a year in advance of setting and to mix well-rotted manure with the soil where the trees were put. the land selected has an eastern, southern and western exposure. The trees began to bear at about twelve years of age, with but a part of the orchard always under cultivation. Portions cultivated one year were left in grass the next, and later cultivated again. This was carried on for 25

> years, when the entire and used for orchard was seeded and used for mowing and sheep pasture. Until the trees were 20 years old the crop was 500 to 1000 barrels annually. Then the grass was partly smothered by a good application of manure, when the yield was suddenly increased very largely.[14]

For many years, however, the orchard struggled. In one of Mary Will's letters (January, 1895), she bemoaned the poor production: "Our apples were a failure this year as usual, what there were tooted badly."

Jonathan raised various fruits to help provide a financial cushion for down-years with the orchard. Figure 5.4 shows the two—fruits and apples—side-by-side.

*Figure 5.4 Fruits and apples*

Things began to change around 1900. First, after years of spraying the trees with a hand pump, a physically exhausting task, the brothers investigated mechanical sprayers.

Looking back, C.T. remembered it this way:

> I heard quite a good deal about the Western New York orchards, and took a trip out there one spring and looked over their power sprayers. I bought one, and I suppose it was the first power sprayer in the state of Vermont. A lot of wise heads said what a

> foolish man Mr. Holmes was to get a power sprayer. We bought that first power sprayer about 15–16 years ago. These wise heads thought it was foolish to put any more into an orchard that never paid anything and never would.
>
> When we think of it, pumping water from a hand pump for 100 acres of orchard trees that were 25 or 30 years old, you can imagine this was something, but we did it until they got up these rapid gear sprayers, and my brother and I fixed up one of those and it worked very nicely and was a great improvement over the hand pump. We used it until it was worn out.

The picture below (Figure 5.5) shows C.T.'s son, Robert, operating the power sprayer.

*Figure 5.5 Spraying*

The following (Figure 5.6) is another view of the sprayer in action.

*Figure 5.6 Spraying*

According to C.T., the power sprayer was important, but not sufficient for ensuring healthy apples. The trees required serious cultivation.

> After I got the power sprayer I made up my mind something else must be done. The trees would blossom but not bear fruit and the little apples would drop off. (Since 1900) I have kept it cultivated and fertilized and calculate to keep it cultivated to the extent of the branches. The trees are 30–35 feet and take up that distance from one side of the row to the other. I had four very good crops of apples from that time, and I had that 7,000-barrel drop, 1900–1904.[15]

The farm had hired hands to help with the orchard, and here (Figure 5.7) is a helper testing the irrigation flow from the lake to the orchard.

*Figure 5.7 Testing irrigation pipe*

Packing the apples was a critical step in the chain of events leading to market. Figure 5.8 shows the orchard team packing the stave barrels.

*Figure 5.8 Packing stave barrels*

Here (Figure 5.9), another team is loading apples into barrels.

*Figure 5.9 Packing stave barrels*

Below (Figure 5.10) is Robert driving a carriage loaded with packed apple barrels.

*Figure 5.10 Barrels on wagon*

The farm had hundreds of the barrels for packing the apples. A barrel, with C.T.'s name affixed, was kept over the years in an old shed at the house in Middlebury where the family moved after the demise of the farm. The barrel was piled among old tools and boxes; throw-away "junk" of the family's past. The significance of the barrel was not realized until 2021 when C.T.'s great-granddaughter saw it, turned it over, and saw C.T.'s name. The barrel, Figure 5.11, is in remarkably good shape more than 100 years after its was constructed. Two wires encircled the barrel to keep the staves in place.

*Figure 5.11 Stave barrel*

A related question is whether the barrels were constructed at the farm or whether they out-sourced this task. Available records do not offer a definitive answer to the question. Because they had a building that contained a machine shop, it is likely that the Holmeses had the skill-set and motivation to build the barrels at the farm. We can assume this would have been a more cost-effective approach than paying someone else for the barrels.

The proper packing of the stave barrels and shipping by rail was an important element in avoiding damage to the perishable product. In 1919 the Rutland Railroad issued instructions on how to properly load apples on a railway car (Figure 5.12; next page), entitled "CROSSWISE OFFSET LOADING SYSTEM, FOR

STANDARD STAVE BARRELS."[16] With all the effort to produce apples for market, no one wanted to make a costly error in the final stage. Loading the stave barrels was an important task.[17]

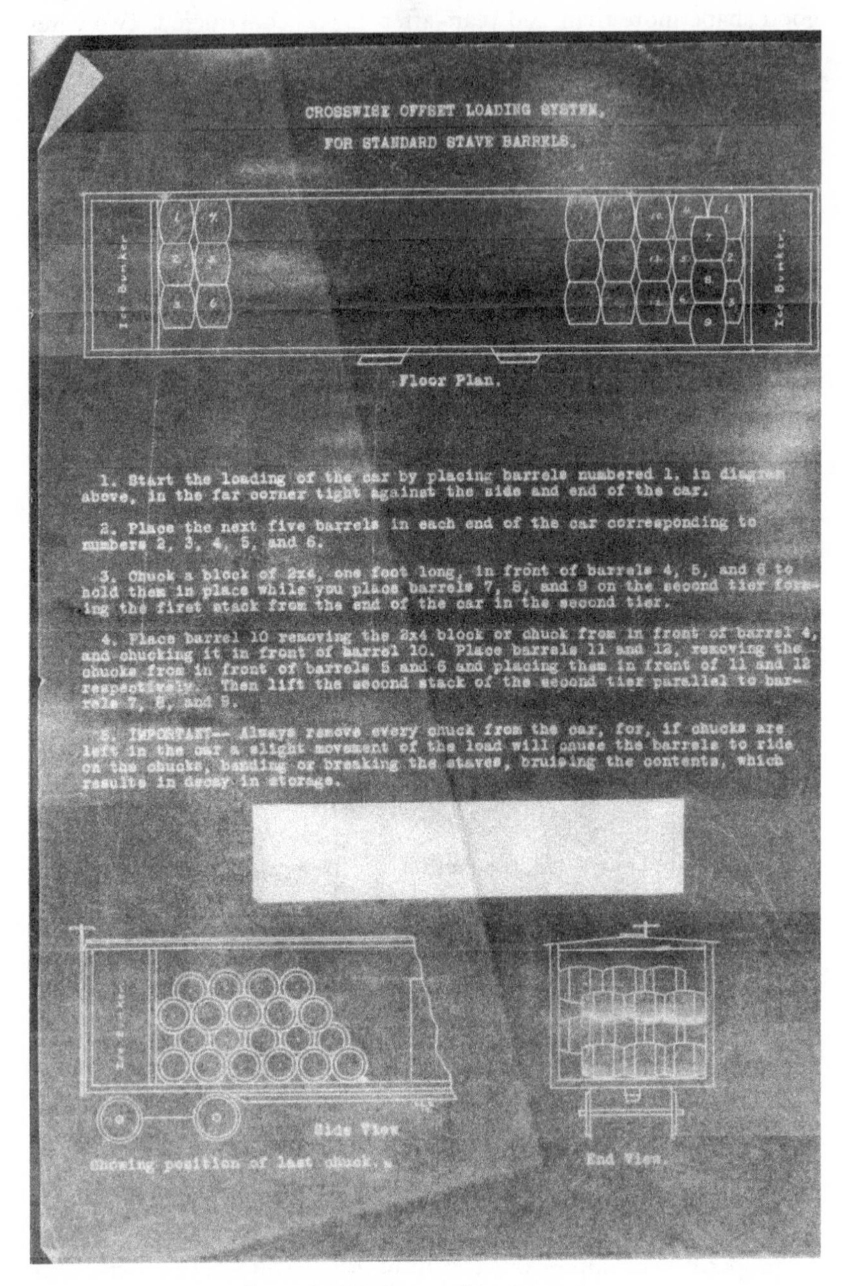

CROSSWISE OFFSET LOADING SYSTEM,

FOR STANDARD STAVE BARRELS.

Ice Bunker

Ice Bunker.

Floor Plan.

1. Start the loading of the car by placing barrels numbered 1, in diagram above, in the far corner tight against the side and end of the car.

2. Place the next five barrels in each end of the car corresponding to numbers 2, 3, 4, 5, and 6.

3. Chuck a block of 2x4, one foot long, in front of barrels 4, 5, and 6 to hold them in place while you place barrels 7, 8, and 9 on the second tier forming the first stack from the end of the car in the second tier.

4. Place barrel 10 removing the 2x4 block or chuck from in front of barrel 4, and chucking it in front of barrel 10. Place barrels 11 and 12, removing the chucks from in front of barrels 5 and 6 and placing them in front of 11 and 12 respectively. Then lift the second stack of the second tier parallel to barrels 7, 8, and 9.

5. IMPORTANT-- Always remove every chuck from the car, for, if chucks are left in the car a slight movement of the load will cause the barrels to ride on the chucks, bending or breaking the staves, bruising the contents, which results in decay in storage.

Ice Bunker

Side View

Showing position of last chuck.

End View.

*Figure 5.12 Railway packing instructions*

We don't know how many, if any, of these well-built barrels were returned to the Holmes farm after delivery to distant locales, both domestic and overseas. If they stayed at the sales destination, it is likely that the cost of the barrels was factored into the sales price for the shipment. Moreover, it is difficult to imagine rail cars or ships returning from a destination with empty barrels for return to the Holmeses.

For the Holmes farm, loading the barrels correctly was an important task. Nevertheless, bruising the apples at the picking, loading and transportation stages was a constant problem and a source of lost revenue.

Around 1900, according C.T., there was a crisis over the ownership of the farm: "By that time there was a lawsuit came up in regards to the farm and it was a question whether to get off the farm or do something and redeem it. We went into the suit and I won out and got the place."[18] Sixty-eight years after Nicholas moved the family to the lake, it appears there was a serious possibility that they would have to abandon the farm. Available records do not reveal what the litigation was about or who was seeking to gain ownership. Keeping the farm, however, was a turning point in building a successful orchard.

C.T. turned outside the family for help:

> After I got the farm I didn't know what to do with it. I had no money to run it, but I finally got some friends interested and got a little cash and a couple of good teams and put in 250 tons of manure and plowed it under and kept that orchard cultivated that season. That fall I harvested about 7,000 barrels of apples. That was a surprise to everybody. That was the turning point for the orchard.[19]

We don't know who was recruited to invest in the orchard or what the financial arrangements were. Taking on loans, however, was a pattern that led to a disastrous outcome.

## C. Thriving Orchard and Growing Recognition

Vermont newspapers and various journals provided extensive coverage of Vermont agriculture, including the apple industry. In the years after 1904, the Holmes orchard began to flourish and attain prominence in Vermont and beyond. In addition to Vermont State publications, the orchard and C.T. were featured in publications such as the Burlington Free Press, the Bennington Evening Banner, the Middlebury Register, and the Barre Daily Times.[20]

Robert, C.T., and William were the family leadership team, as shown here (Figure 5.13), with C.T. in front flanked by Robert to his right and the bearded William to his left.

*Figure 5.13 Robert, C.T., and William*

When I showed this picture of C.T. to my son (C.T.'s great-great grandson), he commented, "He looks like a cool guy." Indeed, the casual stance, impressive mustache, and holstered ax convey an intriguing personage.

The Garden Magazine article of 1911 reported on the stature of the orchard:

> A single orchard in Vermont has advertised the state before the world, as one where fortunes may be made in apples. Its crops of 1900, as sold, realized $20,000 and crops of previous years also brought large sums. Its area is only 100 acres on Lake Champlain, a portion of the farm of Charles T. Holmes, in whose family it has been owned for generations.
>
> Eye openers of this kind occasionally awaken Eastern farmers to the fact that apples may be grown to great perfection and profit here as well as in Oregon and Washington. Our great mistake is this: We do not persistently endeavor to grow faultless fruit in quantity. On Mr. Holmes' farm it is the rule to attend to the requirements of each tree individually, to manure the ground and cultivate it, to prune the tree and spray it as if that were the only tree on the farm.[21]

Once the orchard was fully developed, harvest time brought almost 100 temporary hired hands to help with the picking and packing. A large dorm provided sleeping quarters on the property. Figure 5.14 shows pickers at harvest time.

*Figure 5.14 Pickers*

The picture seems to show two or three dark-skinned workers likely hired from the several Black families that lived in Charlotte, Ferrisburgh, and Hinesburg.

C.T.'s commentary in various publications tells the orchard's story. In fact, C.T.'s commentary constitutes a history of the orchard and, by extension, a history of apple orchards in Vermont. In 1909 the Horticultural Society included an article by C.T., entitled "My Experience with a Productive and Unproductive Orchard." He wrote:

> In the winter of 1907, 50 acres of greenings were given a good mulch of barnyard manure as far as the branches extended. As soon as frost was out of the ground I turned this mulch under about three inches – not deep enough to injure roots. I don't believe in pruning a tree at both ends. One acre was given a dressing of air slacked lime—about 200 pounds to each tree. I will here say that this acre showed marked superiority to the rest in color and size of foliage and in finish and texture of fruit that 70 acres more were limed that fall and the remainder of the orchard will be limed this winter.
>
> To go back to 1907, the plowed ground was thoroughly pulverized with a disc harrow and was gone over with a spring tooth harrow about every two weeks until the middle of July, when I sowed one bushel of buckwheat to the acre. Two weeks before apple picking time, when the buckwheat was in full bloom, it was rolled with a low roller in order to break it down and at the same time provide a soft cushion for windfalls and keep them clean. That fall I picked 2,500 barrels from the 50 acres I took care of, and 600 barrels from the rest.[22]

In the spring of 1908, C.T. listened to others and tried an experiment.

> I foolishly allowed myself to be influenced against my own judgment by the opinions of some of the wise ones, who said the crop of the previous year was due to manure alone. I dropped one-half of the 50 acres and cultivated but 25, while manuring the entire orchard. We all know what a dry season we had in 1908. I cultivated the 25 acres once every ten days until the middle of July when I seeded mammoth clover for a cover crop for winter. The outcome of this experiment was that from 25 acres cultivated two years in succession 600 barrels were picked. The 25 acres cultivated but one year yielded 200 barrels, much inferior in size; and from the other half of the orchard, I got nothing.[23]

For the first time, it appears that the orchard was receiving widespread recognition.

> I was now convinced that intensive cultivation was of the greatest importance, and this year (1909) the whole orchard was under cultivation. From inquiries coming from all parts of the continent, it would seem that everyone had heard of the crop of apples which rewarded me in 1909.[24]

C.T. went on to describe what the apples earned in the market:

> The windfalls brought $2 per barrel at the station. The others were placed in cold storage at Troy, NY, and are being handled by E.P. Loomis & Co., of New York. Fancies are selling at $7 and firsts at $5. Number twos were sold on arrival at $3. The 1907 crop netted, after deducting costs of picking, freight and commission, $3,044.50. The crop of 1908 netted $2,000. This year (1909) the net returns will go well beyond $10,000.[25]

By this time, markets for Holmes apples included Vermont and New York by rail (the Rutland and Burlington Railroad was built through Charlotte between 1848 and 1849) and London by ship where they had an agent to sell the apples.

Figure 5.15 shows the location of the Holmes dock off the northwest shore of the farm. The London-bound apples went north on Lake Champlain to the St. Lawrence River then across the Atlantic.

*Figure 5.15 Apple dock location*

Because of neglect and the pounding of the waves, what residents knew as the "apple dock" eventually fell into disrepair and sank into the lake. Sky Thurber, whose parents took over the farm in 1936, remembers as a youth diving down to inspect the old pilings.[26] Later owners of the shore property built a new dock at the same location.

The critical role of spraying was evident to C.T.:

> In spraying I use a two and one-third horsepower gasoline sprayer made by the Field Force Pump Co, Elmira, New York. A man makes no mistake when

> he buys one of these. Hand power pumps are too laborious and do not give force enough to produce the fine spray necessary to reach every part of fruit and foliage.
>
> I spray before buds open for bud moth and cigar case bearer with Bordeaux-lead-arsenate mixture, and again, after the blossoms fall, for coddling moth, with lead arsenate and lime. My experience is that this is the most important spray. The third spray comes just before the apple turns down, and is for blight, fungus, and the coddling moth. Spraying at this time, if carefully done, fills the calyx with a first and last breakfast for the grub when it emerges from the egg. Our greatest pest is the coddling moth, but this little chap need not be feared if the spray nozzle is pointed in the right direction and held there long enough during the second spraying.
>
> With good cultivation, thorough spraying, trimming, and fertilization, we have nothing to fear from the great northwest. Last but not least, pack true to mark.[27]

What we see again from C.T.'s testimony is an analytical and experimental mind. Although he was not educated beyond grade school, C.T. was a scientist at heart. He tested alternatives, studied the results, and reported to an audience of orchardists.

Vermont suffered four unusually dry seasons after C.T.'s breakthrough in 1909, which affected the production of orchards statewide. Still, C.T.'s reputation and that of his orchard continued to grow. After C.T. was named to the Executive Committee of the Horticultural Society in 1909, public recognition reached its zenith in 1911. The horticultural society had the idea of an annual outdoor meeting and decided to hold its second summer meeting at the Holmes farm on Aug. 24, with this agenda:

> *Opening remarks:* Pres. G.W. Perry
> *Report of the Orchard:* Prof. Wilfred Phelps
> *Addresses (5)*
> *Demonstration of Gasoline Power Engine:* C.T. Holmes & Son

The proceedings of the meeting included the following:

> On arriving in Charlotte, the guests were met at the station by teams and driven to the farm of Mr. and Mrs. Holmes, where the remainder of the forenoon was spent in looking over the orchard and witnessing a free demonstration of dynamite blasting of tree stumps, stones and compact soil. Following the demonstration of Mr. Nichols, a basket luncheon, supplemented by the hot coffee furnished by Mr. and Holmes, was enjoyed by all.[28]

The afternoon program included a demonstration of the gasoline power sprayer by C.T. and his son Robert.

The president of the society, G.W. Perry, opened the meeting by expressing excitement about the potential of the apple industry in Vermont, heralding the role of the Horticultural Society, and citing the Holmes orchard as an example of what could be done:

> We have been visiting today what I think is perhaps the best orchard, taking it all together, in New England—certainly the best very large orchard in Vermont so far as I know.
>
> We are in the valley best adapted to fruit growing—not the only valley, for the valley of the Connecticut is also a fine growing part of Vermont, but this valley (Champlain Valley) is wonderfully adapted to fruit growing, and around this lake everywhere are many acres, thousands upon

> thousands, of fruit land which are yet unoccupied, on which man are raising crops of hay and grain.
>
> The time is coming within the lifetime of these boys before me when every acre will be worth $500. Our men are going into Oregon and paying $300 and $400 an acre for land that will not produce apples as this land here, and we are right at the markets. Our apples can be grown equally as well, and we are least $2 a barrel better off in freight rates that they are out there. It costs them 75 cents a box to send their apples into the East and sell them, and we can ship them from here for about 25 cents to Boston and New York, and we haven't got to pat a man down there to sell them because he will come up here and buy them.
>
> If I were a young man of fair ability and getting an education, I would make apples growing the business of my life, and in ten years become the best posted man on apples in the country.

Perry went on to extol the role of the Horticultural Society.

> A good deal of the present knowledge and interest in apple culture is owing to this society. We have 350 people present here today and out of that number I presume not over 50 or 60 are members of the Vermont State Horticultural Society, which is doing this work and undertaking to wake up our people to their opportunities. This society is growing and becoming a power in the state. If you want to help along this work the best thing you can do is to become a member of the society today and pay in your 50 cents a year and help boom the thing. What a marvelous thing it would be if we could look across these ridges and see every bit of them covered with orchards like this one of Mr. Holmes.[29]

In the annual society publication, the secretary reported, "The annual summer meeting surpassed all expectations in attendance and interest. The total attendance was nearly 400."[30]

The report included pictures of the Holmes farm.[31] In the same publication, it was reported:

> The season of 1911 was an off-year for the fruit growers of Chittenden County. More spraying was done this year than previously, many using lime and Sulphur spray. Some good crops were grown by Mr. Holmes, Dr. Webb, and Mr. Drew. The fruit is not as large as usual, owing to the dry weather.[32]

Also in 1911, the *Vermont Agricultural Report* included an article by the previously mentioned Professor Cummings of the University of Vermont Agricultural College. Cummings, a descendent of early Vermont settlers in Thetford, was a professor of horticulture at the University and secretary of the horticultural society for 30 years. He followed closely developments at the Holmes orchard and, like C.T., experimented with apple growing.

Years later, in 1937, Cummings discovered the emergence date of the fruit, which gave farmers useful data on when to spray to keep fruit flies under control. His 1911 article, entitled "Apple Culture in Vermont," gave an overview of the apple industry nationally, citing statistics that indicated the "rapid decline in the apple production."[33]

Despite his sobering report about the national picture, Cummings went on to describe the great potential in Vermont, noting that "Vermont is a natural apple state" and "to become familiar, experienced and successful in apple orchard culture a man must study his business, carefully and thoroughly."[34] The latter statement describes what C.T. brought to his orchard, and Cummings noted his successes:

> Some of the people in Vermont have already realized our possibilities in growing. C.T. Holmes, of Charlotte, perhaps known to some of you, has had remarkable success with apples in the last few years.

> Last year, 1910, his crop gave him $20,000 from 100 acres. Taking out the expense for maintenance he has left nearly $15,000 for the year's crop. The last year before that he cleared about $12,000, and the year before that somewhere about $7,000. There are other men doing perhaps as well according to the acreage and whose income is great.[35]

After describing the necessary steps to create a successful orchard, with an emphasis on spraying and pruning, Cummings ended on an optimistic note:

> I want to say in closing that the apple is one of our great staple crops. The apple should be to Vermont what corn is to Iowa and cotton is to Georgia. I have heard people say you cannot grow corn in Vermont, but I know you can. I have heard people say you cannot grow apples in Vermont, but I know you can. I have great faith in our possibilities.[36]

As noted, the Vermont Commissioner of Agriculture issued a 31-page overview of the apple industry, *Vermont, An Apple Growing State,* which was written by the omnipresent Cummings. At this time in the early 1900s, the report was a comprehensive primer on the state-of-the-art in developing a successful orchard. The report referred to "The largest orchard in the state," owned by C.T. Holmes of Charlotte.[37]

The January 1911 article in *The Garden Magazine* noted that someone tried to buy C.T.'s orchard and farm:

> The outlook is so good that that the painstaking owner has been offered $50,000 for the place. And after all, the soil is ordinary New England land. The trees are what enhance its value. The price offered Mr. Holmes, $50,000, would seem to everyday

> farmers to be extravagant for 100 acres of orchard, but upon analyzing the purchase it is very modest.[38]

C.T. next appeared in the Horticultural Society's annual report in 1914. He made remarks at the annual summer meeting and responded to questions from the audience. The article that appeared in the journal was entitled, "Managing 100 Acres of Large Trees."

The first signs of trouble for the orchard surfaced in C.T.'s commentary:

> I have kept it cultivated and fertilized and calculate to keep it cultivated out to the extent of the branches. The trees are 30–35 feet and take up that distance from one side of the row to the other. I had four very good crops of apples from the time I had that 7,000-barrel drop, and since then I have noticed that there has been a deterioration in the quantity and size of the apples. I attributed that to the first of few dry seasons we had four years ago, a condition which has been growing worse every year until this year. It is frightful the amount of dead wood I have taken out.
>
> It is a question of what to do. I am studying up now an irrigation problem, for I think that is the cause of the whole trouble, especially those punky brown spots in the apples. I noticed them first about three years ago, and this year they were all over; worse in some sections than others. There was a party at my place yesterday talking about this irrigation problem, and if everything is right, I shall have a stream of water on top of that ridge with an 8-inch main, with laterals to different sections of the orchard and a pump with a capacity of 700 to 800 a minute.

In answer to the question "Do you expect to get a fair crop every year you irrigate?" C.T exuded optimism.

> I do sir. At the time I got that manure I put in about 150 pounds of lime to a load. I spread that on first and manure on top of it and turned it under, so that the lime would not work on the manure and injure the value of it. I think I shall start in with that once in about four or five years. I have been all over the orchard once and part of it twice, and this fall, or another fall anyway, I shall re-lime the whole orchard again.

C.T.'s phrase, "I do sir," has a familiar ring. I heard my relatives, including my uncle, Charles Ross, speak in a similar manner in public forums. Charlie Ross ran for the U.S. Senate in 1974, and I saw him in action on the campaign trail. Both C.T. and his grandson, Charlie, responded to questions in a way that implied expertise as well as deference to the audience.

C.T. was asked about his approach to irrigation:

> I have a tank that holds 4,200 gallons at the highest point of my orchard, and will also take it to my son's house and my own. There are two trees that stand next to where we drive our sprayers, and Professor Cummings was down there this summer and was talking about irrigating a couple of trees there, and I said I would do so. The next day I attached a one-half inch hose and dug a little ditch around the trees and let this stream run for three days, and that convinced me that one wetting was what that orchard wanted. The trees in the orchard looked sick and bad, and on these two trees there is not a blighted leaf or dead limb, and the apples were green and nice size, like I used to raise. I shall irrigate by the centipede system or ditching. I took it up in a

> small way for I wanted to know what it was going to cost. The pump and engine will cost something around $150–$180. The piping and outlines are the most expensive.

C.T. went on to discuss the importance of lime, noting: "I think the lime in the soil has gone out and we have to have more."[39]

The 2021 comments of an orchardist in the Champlain Valley suggest that C.T. may have had it wrong in his approach to keeping the trees moist during dry spells.

> To irrigate and cultivate is never going to retain moisture. Then it must have been part of the Ag culture or C.T.'s own experiment. We, along with UVM and Cornell and most others, mulched with hay in dry years to hold moisture or just let grass get long and mowed with a sickle bar to put down a layer.[40]

If, in fact, this method of retaining moisture—laying down hay or grasses—would have worked for C.T., one can imagine the orchard thriving, even during the upheavals of World War I and a run of dry years. The calamity of losing the farm might have been avoided. Whether or not laying down hay or grasses would have worked for C.T.'s trees, the advice comes 100 years too late.

## *Hired Help*

Cameron Clifford's history of farming in North Pomfret makes clear that hired hands were a critical need at various times of the year. Reliability was always a question with the hired help, and Clifford reports the time that "Ed Dana's hired man up and left one day without any notice." When his neighbor, Carl Johnson, heard the news, he went over to help Dana with the milking.[41] A farm owner often decided to keep a good hired hand on during slack times in order to have his help during busier times.

Like other Vermont farms, the Holmeses employed extra help, with substantial numbers required during fall apple picking time. Here is a picture (Figure 5.16) of one of the long-time hired hands.

*Figure 5.16 Hired hand*

The family needed to supervise, feed, house, and pay the workers, a management task that was critical to the success of the farm and orchard. According to one report, "During apple picking time, a crew of 100 men worked from dawn until dark. They ate five meals a day and Mrs. Whalley made the bread to feed the hungry crowd."[42]

Frank Ellison and Henry Claxton, both relatives, were regular helpers during picking season. The next picture (Figure 5.17), taken during apple picking, suggests the manual labor required to get the apples: move the ladder and place against the tree, climb the ladder with the basket, pick the apples within reach, climb back down the ladder with filled basket, repeat. C.T.'s full-bearded brother, William, is at right.

*Figure 5.17 Pickers*

C.T., Robert, and William were hands-on managers and worked closely with the hired hands in the orchard.

The farm property had a boarding house and two buildings designated as worker housing. We know little of the living conditions for these temporary workers or what kind of managers the Holmeses were. For sure, the work was physical from dawn to dusk. Also, we have minimal documentation on where the hired hands came from to help with the apple picking. At the time, however, political leaders were sensitive to agricultural entities hiring "aliens, any foreigner, or foreigners," which was formalized in congressional legislation in 1885 and later. Xenophobia, which was widespread in America at the time, played a role in these attitudes.

An incident at the farm in 1910 involving workers provides insight into the challenges of hiring and supervising temporary workers. A *Burlington Free Press* article on Sept. 24, the height of the apple season, had the headline:

**APPLE PICKERS ARRESTED**

**Got drunk and raised a rumpus at C. T. Holmes's house**

**Sheriff Allen and deputies made quick automobile trip to Shelburne where they found the men on a train**

The article told a story about a dance the night before, a night of drinking, and C.T. Holmes firing eight workers because of their intoxication.

> Tuesday night they became intoxicated, it is said, and were discharged by Mr. Holmes.
>
> Yesterday morning the men, eight in all, proceeded to the Holmes premises, where they started in to 'clean up' the place, breaking furniture, and raising rough house generally. After they had done considerable damage the men started for the railroad station. Mr. Holmes immediately telephoned to Chief Allen that the men were on the way to Burlington. All belong here in Burlington except one man.

The article goes on to tell how the sheriff and his deputy raced by car "in a cloud of dust" to Shelburne, located between Charlotte and Burlington, on the chance the men might get off the train there. Two escaped in Shelburne, but the sheriff brought six back to the Burlington county jail.[43] Apart from the "sensational" dash to Shelburne and the subsequent arrests, one can surmise the chaos and fear that must have overtaken the Holmes household. This was a breakdown of law and order on the Holmes farm, but perhaps not so surprising considering how urgent it was to hire "live bodies" to help with picking.

The paper profiled the culprits, including Frank "Jigger" McKenna. McKenna's grandparents came from Ireland to Burlington, and their grandson—Frank—was convicted of highway robbery and released from Vermont State Prison in Windsor 1910, just a few months before the incident at the Holmes farm. A 1911 Free Press article reported Frank's frequent returns to court as a serial intoxicant and suggested his situation was due to "genealogical ramifications." This was a veiled reference to his Irish heritage and supposed propensity toward alcohol.[44]

## *Feeding the Crew*

A farm, like an army, travels on its stomach. With a physically active and thirsty group of workers, providing meals and liquids was a vital job. As noted above, the family fed up to 100 workers during the height of apple picking. Special occasions, such as hosting the Vermont Horticultural Society in 1911, brought an added challenge. The society report of the summer 1911 meeting indicated that the Holmeses provided basket lunches and hot coffee for 350 visitors.

Mrs. Whalley, who made bread for the crew, was a relative by marriage and lived about a mile south of the farm. In addition to Whalley, the Holmes women at the farm joined in to meet the many demands of apple-picking time. There must have been abundant kitchen and cooking equipment to provide food for so many mouths, but we lack pictures or documents on this dimension of farm life

## *When Things Were Good*

Luther Putnam, president of the Horticultural Society, closed the 1911 meeting at the Holmes farm with a comment about the expertise of Vermont orchardists like C.T.:

> I just want to say a word, and I think every person here will bear me out, that we do know what the value of these witnesses before use are today. You don't need to send off to Massachusetts or any other place so far as these witnesses go. These men that come from away are necessary, and we want them,

> but when we have these living witnesses right at our door, we don't know what we miss when we are not here to hear them.[45]

On that August day in Charlotte by the lake, C.T. Holmes was a "living witness" to the emergence of the Vermont apple industry. The orchard was an accomplishment of great pride for the family and brought recognition beyond the environs of Charlotte. A lasting memento of the Holmes orchard is a copper plate (Figure 5.18) affixed to every outgoing stave barrel.

*Figure 5.18 Copper plate*

The prominence of the orchard signified the Holmes family's successful journey from England to colonial New England, to the Vermont frontier in the late 1700s, and to the farm along the shores of Lake Champlain that lasted for four generations.

From 1890, the hard work and inspired entrepreneurship of the family began to pay off with success in the farm's various enterprises, including breeding and racing trotters, building a flourishing apple orchard, and working the land to grow several agricultural products. They overcame the dangers faced by all farmers (volatile markets, bad weather, accidents, illness, managing cash flow from planting to harvest, etc.). Things were looking up. Dressed up for a family portrait in the late 1890s (Figure 5.19), C.T.'s family looked the part.

*Figure 5.19 C.T.'s family*

The picture shows C.T.'s family: Clara and C.T., with the children, from left, Robert, Hannah, Maude, and Mildred. Sadly, young Mildred died a short time after this picture was taken.

## Chapter Six

# Foreclosure and Moving On

*The orchard was the family's defining accomplishment. When the orchard came on hard times, the whole enterprise—farm and orchard—were put in jeopardy.*

The Holmes family came on hard times after 1915. World War II, the Spanish Flu pandemic, and growing economic pressures on the orchard combined to alter the life of the family. The story is one of hardship as well as resilience and recovery.

## A. The Great War and the Spanish Flu

Dr. Jacob Ross was married at the Holmes farm in Charlotte to Robert's sister, Hannah, on June 9, 1909. A few years later, when Jacob moved his medical practice from Richmond to Middlebury, he apparently faced resistance. In an interview with the author in 1985, Jacob's daughter, Helen, reported that the lone doctor in Middlebury resisted Jacob's move to the town and bought up all the serum to combat the Spanish flu. Fortunately, a Middlebury College professor and another citizen petitioned the State Medical Board, and Jacob received an allocation.[1]

It is impossible to confirm the accuracy of Helen's depiction of internecine competition among the physicians. We know, however, that the serum situation in 1918 was very different from the rollout of vaccines to combat COVID-19 in 2020–2021. As one writer observed:

> It was a free-for-all. A wave of independently developed vaccines were distributed throughout the country by doctors and researchers who were confident they'd found the answer. One problem: The vaccines developed at the time were ineffective, largely due to the fact that the flu was mistakenly believed to be bacterial in nature. It wasn't until the late 1930s, when scientists determined the cause of the flu was viral that the first effective flu vaccines were produced.[2]

The irony of any effort to deny Ross access to the serum was that the serum was not effective anyway. Serum or not, Jacob set up his practice, visiting his patients throughout Addison County.

Influenced by the Morgan horse legacy of the Holmes farm, Jacob kept a Morgan horse in his barn adjacent to his house in Middlebury. Knowing how close Hannah and Jacob were to the Charlotte Holmeses, it is likely that Jacob's horse came from the stable at farm.

Pulled by his Morgan horse, Jacob traveled by carriage to his patients as far away as Ripton in the mountains southeast of Middlebury. His son, Charlie, recounted traveling with Jacob on visits to patients. Jacob's life as a town doctor came to a halt, however, when he volunteered for the medical corps and traveled to Georgia in April 1918 for training before heading to the European front to serve as a physician.

As always, illness and disease were a fact of life, even more so when the Spanish flu spread to Vermont in the fall of 1918. On Oct. 4, 1918, Vermont state officials closed schools and banned public gatherings. The pandemic was particularly deadly in Montpelier and Barre, with the largest loss of life in Barre where there were almost 200 deaths. By the end of December, there were 43,735 confirmed cases and 1,772 deaths in Vermont.[3]

Looking back, historians have observed that the Spanish flu epidemic brought a "mass amnesia" about its devastation and has been forgotten in the retelling of the times. Historians say the pandemic sank into oblivion largely because of World War I, the very cataclysm that hastened the spread of the virus via millions of moving troops. The war and its aftermath overshadowed the disease too. For the Allies, when Armistice Day came in November 1918, there was a victory to celebrate.[4]

Catherine Arnold, author of *Pandemic 1918: Eyewitness Accounts from the Greatest Medical Holocaust in Modern History*, posited another reason for the collective forgetting:

> Part of the problem was that dying from the flu was considered unmanly. To die in a firefight, that reflected well on your family. But to die in a hospital bed, turning blue, puking, beset by diarrhea—that was difficult for loved ones to accept. There was a mass decision to forget.[5]

For Vermonters in 1918, however, the flu was very real and, similarly to COVID-19, loomed over Vermonters. For the Holmes family, the flu cast a shadow, and they prayed that it would not take one of their own.

The double-trauma of a devastating pandemic and a world war touched families across Vermont. The absence from home of Jacob weighed heavily on his wife Hannah and the extended family. We know about these years because of a treasure trove of Jacob's letters home and Elizabeth's letters to Jacob. Jacob and Hannah's letters were saved and are preserved in the Henry Sheldon Museum of Vermont History in Middlebury.

Despite having to leave his family, Jacob was proud of his decision to join the war effort. In May 1918 he wrote Hannah that "I would consider myself a coward not to go."

He traveled on the SS Aquitania to England, then arrived in France in June 2018. Along with descriptions of the war, attending to injured soldiers and his daily life at the front, Jacob's letters dealt with mundane family matters, such as getting money to Hannah, arranging rent payments on land in Huntington, the health of his children and extended family, and staying close to the Holmeses at the farm in Charlotte. The image below (Figure 6.1) shows Jacob somewhere in France dressed to withstand the cold, about to embark on a flight.

*Figure 6.1 Jacob Ross in France*

Hannah's letters to Jacob are particularly valuable. Providing insight into the war years, she wrote 18 letters over seven months (Aug. 2, 1918 – March 1, 1919), and they offer a look into the daily life of the family, both at the farm and in Middlebury. The letters reveal several dimensions of life in 1918 and 1919: love and concern for each other in the family, worry about those serving in Europe, anxiety about money, and fear of contracting the Spanish flu. Hannah grew up on the farm and now lived with her three children in Middlebury, just 25 miles away.

Her letters show that she stayed close to her family at the farm:

Mother's being here since Thursday. Friday we visited all the evening and Saturday. Folks are busy haying (at the farm) and then comes harvesting soon. Mother (Clara) is talking now of wanting to come home (to the farm). Wish she would stay longer (in Middlebury) as I think it would do her good.
*–Aug. 2, 1918*

Father (C.T.) is busy with apple picking. Expect about 2,000 barrels of firm apples. Have about 20 men. No trouble getting help though they feared it.
– *Oct. 4*

I did want to drive up home (the farm) so bad but knew I shouldn't after washing so I told K (her daughter, Katherine) if it was good day we'd go Tuesday, yesterday morning. It was quite cold but about 9:30 the sun appeared. I took a look at the old car and she was OK, so I told K we'd get ready. In just 35 minutes I had our clothes all changed, hair combed and wraps and at 10:15 I cranked her on the sixth crank and we never stopped except to fix a side curtain till we struck home in just one hour and 35 minutes. There were some pleased to see us. I guess I am so worried over father (C.T.). He is working as you can imagine. They finish apple picking tomorrow (about 1,100 bbl. of apples he thought) and the men stay on one day to pull because his potato crop is rotting about half, due to not being dug earlier, he says, yet he is the same brave man. Says he has not time to be sick. Mrs. Duclos has the men at the boarding this week. It was fortunate she didn't have the flu.
– *Oct. 23*

Father is on jury this week and I expect mother down (to Middlebury) for a few days. Hope it remains warm as she requires so much heat and I'm not running the furnace heavy. Am hoping to make this coal last.
*– Dec. 8*

Mother comes tonight for a few days, the first time since this fall. Hope father can get down for one. He is on jury.
*– Dec. 12*

I did not get a letter off this week owing to went to the movies. Father (C.T.) and Marion (Elizabeth's niece and Robert's daughter) walked in yesterday. Was expecting father last night but did not expect Marion. Was so glad he came. He fixed a piece of soldering on my washer and the pounder. How glad I'll be when I have my own big man to do it for me. Daddy has written you two lovely letters: one, I know Mother got the address all mixed up. Mother is suffering terribly with that numbness in her hands, otherwise pretty well. Father has spent the day reading "Winged War Fare." Great book!
*– Dec. 15*

They have butchered at home (the farm) and roasted pork, some liver and sausage came down today. Wish you were going to enjoy it.
*– Dec. 21*

"It was a great surprise to me to hear Thursday that father had business in Middlebury and he and mother would be down Saturday noon for this Sunday. He hopes to place a big power sprayer for elm trees here and if so will make some $300 on

> it. He thinks it looks favorable. He seems to enjoy being here and I surely like to have him come. While we were visiting last night he said he wished Jacob was here. I guess were all wishing it. Father is looking well. Guess his hernia bothers a good deal but he never says anything and I can't get much out of him. Rena and Robert think he ought to go the hospital at once.
> *– Feb. 16*

Christmas was a difficult time for Hannah, but she had the farm and family nearby.

> The days are fast approaching Christmas. How we miss you—two years without you at this season—but you have been doing your part to help this old world to a happy peaceful Christmas, that is sure.
> *– Dec. 21*

> Maude (Hannah's sister) and Katherine started for Charlotte (the farm) at 5 tonight. We go with babies tomorrow. I can tell you more about Christmas when we get home. How I hope I can tell it in your ears. On my dear, do you guess how much I love you. I am so glad my love has meant a little to you over there, and am more glad than words I can tell you, you have been a help to the boys in keeping them from temptation. Bless you, dear, and grant we may spend next Christmas together . . . How much we have to be thankful for compared to homes so saddened and suffering from war.
> *– Christmas Eve*

Hannah's letters also addressed concerns about other members of the family, the vagaries of the weather, raising her children, managing money, her church, train travel between Middlebury and

Burlington, and the family's patriotism in a time of war. Her reporting on sickness and the Spanish flu, however, concerned friends and family everywhere, including the farm. The situation is eerily similar to COVID-19 a century later.

> This a lousy time for doctors, though the worst is over here in Middlebury, at least as far as college (Middlebury College) goes. Colds, grip, influenza, schools and churches and movies are closed and everybody is staying home. Professor Robinson's wife has died and blessed relief to her and her family. She was entirely out of head at the last. I wrote you Mrs. Skilling died.
> – *Oct. 4*
>
> Burlington is sending doctors to Montpelier and Barre where there are 3,500 cases of influenza. Nurses are at a premium and soon they won't be fussy about an applicant having a high school education. A letter from mother just today says there were lots of cases of sickness in Charlotte. I do hope they (the family) escaped, as they are so far from a doctor.
> – *Oct. 6*
>
> Dr. Norton had told them there was no danger for Robert (Hannah's brother) last Saturday and we went up to see him. Poor boy, he certainly had had a siege, so thin and weak. Schools will probably open Nov. 4.
> – *Oct. 23*
>
> Professor Wright is ill and away and Dr. Collins is acting president (of Middlebury College). Chicken pox is appearing every now and then. I

> had Doctor Flagg came in once to see Ruth and he appeared petty tired.
> *– Dec. 15*
>
> Eunice's people have the flu and she is pretty anxious to get there. They telegraphed her Saturday that Maggie was just alive.
> *– Dec. 30*
>
> Eunice stayed on this week for hoping to hear from home. Friday she heard Maggie had died the same night they sent a telegram. Saw Professor Cady on my way to Sunday school and he asked what you were going to do (when Jacob returns from the war) and I told him help take care of flu, I guess. He said, 'I guess he'll find plenty of it.' Is raging in Bristol and in some other towns.
> *– Jan. 1*
>
> Just heard this morning Rachel was dead, flu, quite suddenly I guess. Poor cousin Ann. It is one more blow.
> *– Feb. 11*

Fortunately for the Holmes family at the farm and the Rosses in Middlebury, no one succumbed to the flu, although young Ruth died of influenza-related symptoms at the age of 5. Robert recovered his strength, and Jacob came home safely in March 1919.

Upon his return, Jacob attained stature in the medical profession, both locally and nationally, and was regular commentator on the challenges of providing medical care in rural communities. An article by Jacob in the February 1920 issue of The Modern Hospital, "How to Meet the Needs of Rural Hospitals," discussed this situation:

> So as we are again approaching the normal there is still a great shortage (of doctors) which will be even

> more keenly felt as the older practitioners retire from service. These men have grown up with the towns until they are a part of the community, and they will spend their last years of usefulness for the "home folks." But we must face the fact that that a man who has to take a two-year course in college, a four-year course in medicine, and a year or two in a hospital as an intern, will be slow to locate in a community of 800–1,000 people.[6]

The article was the print version of a presentation by Jacob at the Second National Conference of Country Life in Chicago on Nov. 8, 1919. Jacob's lament sounds very similar to the situation in today's rural America.

Despite the end of World War I, the travails of the Ross family did not end upon Jacob's return from the European front. Jacob contracted appendicitis and, according to Helen's recollection, was misdiagnosed for three days, and was operated on too late.

Jacob died in 1929, and his death left the family with no money after the debts were settled. Jacob's son, Charlie, remembered his time with his father:

> My father was a country doctor in Middlebury, Vermont. During the twenties, the end of one transportation era, he had two automobiles. But he couldn't depend on them in the mud season or in a snow storm. So had horses. I can remember making calls with him throughout the country before I was in grade school. In a sleigh in the wintertime and in a buggy in the mud season. He enjoyed horses. My father died when I was nine years of age.[7]

Charlie expressed gratitude for what his parents had done for him:

> My real heroes were my parents. They conveyed in me and instilled in me a sense of service to the

> public. In the sense that a person who has some advantages should help his fellow beings.[8]

To compound things for the family, Hannah contracted tuberculosis in both lungs in 1933. According to Helen, she refused to go to the sanitarium in Pittsford (20 miles south of Middlebury), "probably because she couldn't afford it." Hannah was given a year to live but lasted four years and died in 1937. In Helen's memory, "She was okay until year two, then got bad but still came down to dinner, and then was bed-ridden the past two years."

The oldest daughter, Katherine, who was educated at Radcliffe College, came back from California to take over her mother's math teaching job in Middlebury to make money for the family. After the death of both parents, Kay was essentially in charge of the family. With the ongoing TB threat, Helen, the youngest, was sent to Minnesota to go to school.

Hannah's son Charlie had a different trajectory. Jacob had become the physical director at Middlebury College when he moved to town. He succeeded the famous baseball star, Ray Fisher, in this position at the college. Fisher had grown up in Middlebury and went on to be a star pitcher in the major leagues, then became a legendary baseball coach at the University of Michigan.

The Ross family and the Fishers crossed paths in numerous ways. Importantly, the Rosses and the Fishers had summer camps at Long Point on Lake Champlain in Ferrisburgh and saw much of each other. After Jacob died in 1929, Fisher befriended Jacob's son Charlie and took Charlie fishing in the mornings at Long Point. In the afternoons, Fisher and Charlie pitched to each other near Fisher's camp. Under Fisher's tutelage, Charlie became an exceptional baseball pitcher, and starred in baseball as well as ice hockey during his years at Middlebury High School, graduating in 1937, the year his mother died.

When it came time for Charlie to head off to college, Fisher, the Michigan baseball coach, convinced Charlies to attend Michigan. With the Ross family in a dire situation financially, Fisher arranged for several part-time jobs to help Charlie cover the costs of his education.

Charlie starred on the mound for Michigan, and years later he regaled me with stories of facing Hall of Fame pitcher, Robin Roberts.

The Fisher baseball legacy continued when Fisher-trained Charlie worked with his son Peter, who followed in Charlie's footsteps and became a star pitcher at Michigan. As for me, Charlie taught me how to throw a curveball at family picnics on Mt. Philo and Lake Dunmore, which carried me through successes in Little League and high school baseball. Perhaps, most importantly, Fisher introduced Charlie to his future wife, Charlotte, whose father was the Michigan track coach.[9]

Although the Ross children survived, and ultimately thrived, the 1920s and 1930s were a calamitous time for the Rosses of Middlebury. Sadly, both parents died all to soon—in 1929 and 1937, respectively. During the same time, the apple orchard in Charlotte was in a downward spiral.

## B. Demise of the Orchard and Foreclosure

The records of the Vermont State Horticultural Society in 1915 show C.T. served as the county vice president for Chittenden County and on the Society's executive committee.[10]

A footnote to this story is that C.T. knew young George Aiken of Putney, the future politician and legendary U.S. senator. Aiken grew up on a farm and at age 14 joined the Putney grange. After graduating from high school in 1909, he bought 40 acres with a partner and planted fruits, wildflowers, and apples. Over the succeeding years, the Aiken Nursery thrived, and Aiken became active in statewide horticultural matters. The Aiken-Holmes connection came in 1915 when C.T. and Aiken served together on the executive committee of the horticultural society. Aiken and C.T. were together at society meetings and at society-sponsored events. In 1917 at the young age of 25 Aiken was elected president of the society.[11]

The 1915 horticultural society journal reported ominous developments for Vermont apple growers in Chittenden County, the locale of the Holmes farm: "So far as the apple crop is concerned it was as near a failure in Chittenden County as it had ever been. On

Shelburne Farms we had less than 100 barrels, and the wind blew those nearly off, and that was the case throughout the county."[12]

The difficulties facing the Holmes apple business had some ironic aspects. An article by M.B. Cummings in the 1921 edition of the Society's Annual Report, entitled "The Rise of Commercial Apple Orchards in Vermont," included a lengthy paragraph on the "famous orchard of C.T. Holmes of Charlotte" and described the growth of the orchard, including the exceptional yield of 3,707 barrels in 1909. The author embellished an earlier story when he wrote that "a man once offered $55,000 for this orchard property." He added $5,000 to the earlier version. It is telling that this 1921 article extolling the successes of the Holmes orchard did not report up-to-date news about the fate of the orchard.[13] As it turned out, the years after 1909 were not good ones for the orchard, leading ultimately to foreclosure just two years after the Cummings article.

While extolling the Holmes orchard, Cummings had some bad news. He presented data that showed that after 1909 the commercial orchard industry in Vermont went into recession. Unquestionably, World War I and related economic upheaval had a negative effect on production.

The drop in production between 1909 and 1919 is shown below:[14]

| Year | Mature and bearing trees | Too small to bear | Yields in bushels |
|---|---|---|---|
| 1909 | 1,183,529 trees | 219,883 trees | 1,459,689 |
| 1919 | 712,594 trees | 249,029 trees | 960,252 |

Cummings also noted that the figure for bushels in commercial production was significantly lower than the estimated *total* production of apples. He surmised the gap between total production (commercial and non-commercial) and commercial only was due to wastage and home consumption. Cummings asked facetiously, "Do Vermonters eat apples? It would seem so."[15]

The downward trend in apple production in Vermont coincided with a similar trend across New England. A report, *The Apple Situation in*

*New England*, published by the Extension Services of the New England states, documented production for the years 1890 to 1925 (Figure 6.2).[16]

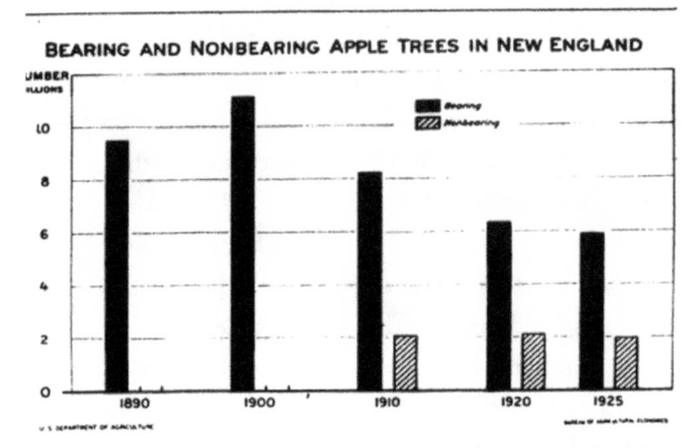


g. 1—The number of bearing apple trees in New England has declined from over eleven million in 1900 to slightly less than six million in 1925.

*Figure 6.2 Apple production chart*

There were four exceptionally dry years after 1911. According to another Cummings article in 1925, these years were followed by the "cold winters of 1917 and 1918 (that) swept out of existence a great many of the small orchards . . . "[17]

As late as 1924, however, the Holmes orchard was listed mistakenly as an example of a large Vermont orchard, pointing out that C.T. Holmes had 2,000 trees on 20 acres.[18] The sad truth was that by 1923 the Holmes orchard was no longer.

Although the Holmes family maintained the usual array of agricultural activities, the orchard had become the driving motivation for C.T. and the family. While they raised, raced, and sold horses, made bricks, raised crops, and kept cows and sheep, the orchard was the family's defining accomplishment and greatest source of recognition and satisfaction. When the orchard came on hard times, the whole enterprise—farm and orchard—were put in jeopardy.

Like farmers everywhere, the finances of the Holmes farm entailed a series of bank loans and repayments. The latter depended on cash income from sales, with the largest income coming from apples. In 1906 C.T. consolidated and simplified ownership of the farm with a transfer of deeds from relatives, including his brother, William. A year later, C.T. and his wife Clara paid off a $5,000 loan to the Burlington Savings Bank.

With the weakening of the apple business after 1910, C.T. took on new bank loans to finance the operations of the orchard, including the innovative irrigation system that drew water from Lake Champlain for the apples and other crops.

Farm finances reached a serious juncture on Jan. 20, 2017, when Charles and Clara executed a mortgage deed for $19,000 with William H. Tinsdale, a resident of Essex County. The mortgage provided cash to pay off debts and needed operating capital. The arrangement with Tinsdale included a series of promissory notes held by the Winooski Savings Bank with these commitments:

$2,000 by June 1, 1922

$2,000 by June 1, 1923

$2,000 by June 1, 1924

$3,000 by June 1, 1925

$10,000 by June 1, 1926

The deed stipulated that the Holmes would keep the farm in good shape and "in husband-like repair." Although the farm operation had many facets, the property description for the mortgage indicated that the apple orchard was considered the farm's primary asset.

> Being the whole of the farm known and called the John Holmes farm, about one hundred acres of which the said John Holmes set to an orchard of apple trees which now exists on said farm. Together with all the buildings and erections of whatever name and nature now or shall hereafter be erected on said farm including the irrigation plant so-called established on said farm for the purpose of taking water from Lake Champlain and bring on the land of said farm. Said irrigation plant consists of a

> forty-horse power Fairbanks Morse engine, water pump and foundations and house thereof and a system of piping and valves to distribute the water over the land of said farm together with the electric lighting plant now established on said farm and in the buildings thereon.

In addition, C.T. and Clara conveyed the farm through a Declaration of Trust to Clarence Ferguson. The trust arrangement provided that C.T. and Clara would be agents and managers thereof for the purpose of carrying on said farm with orchard. They would have use of the houses and farm buildings and be paid wages for managing the farm. The document contained specific expectations for overseeing the farm and for the flow of income and payments.

Essentially, the family had sold the farm to Tinsdale, subject to regaining ownership through payment of the promissory notes: "When the said mortgage to Tisdale is paid in full . . . then this trust shall cease and the said trustee will re-convey to C.T. and Clara Holmes said farm of land and personal property."

Unfortunately, in the years after executing the 1917 mortgage, conditions—national and local—worsened due to World War I, the devastating Spanish flu, and a weakened economy. By the time the first payment was due in June 1922, cash was gone and farm income was well short of what was needed to operate the farm, support the family, and make the June 1922 payment.

After falling short on the first payment, Tisdale and the bank moved to foreclose on the property. The Decree of Foreclosure on May 8, 1923 stated, "A large part of the interest and principal due and payable on said notes remains unpaid and unsatisfied and now justly due and payable." The foreclosure document copied into the records of the Charlotte Town Clerk is written in clear longhand and captures the end-game of the Holmes farm. Here is an excerpt (Figure 6.3):

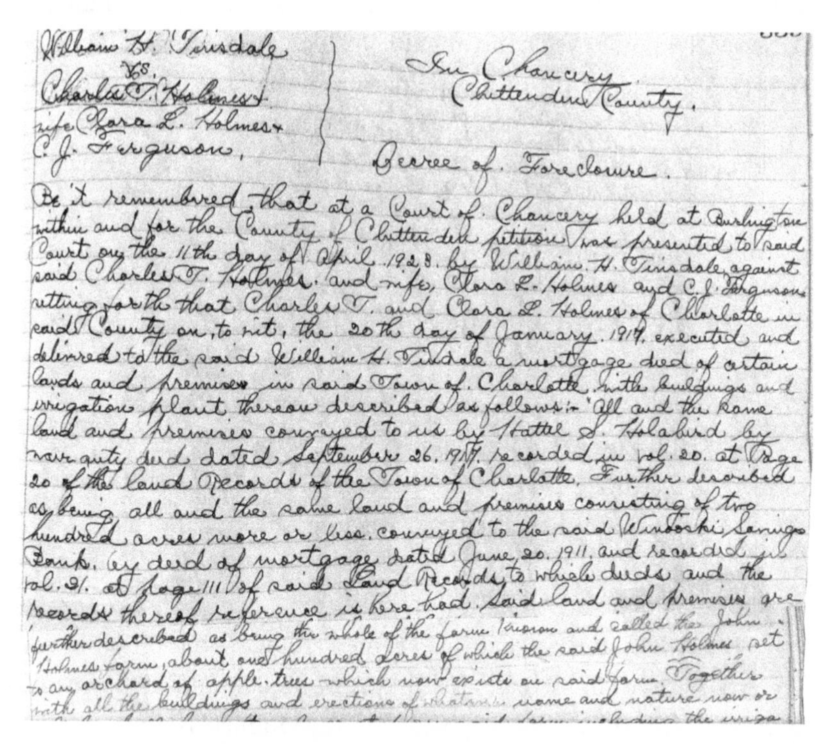

William H. Tinsdale
vs.
Charles T. Holmes &
wife Clara L. Holmes &
C. J. Ferguson,

In Chancery
Chittenden County.

Decree of Foreclosure.

Be it remembered, that at a Court of Chancery held at Burlington within and for the County of Chittenden petition was presented to said Court on the 11th day of April 1923. by William H. Tinsdale against said Charles T. Holmes. and wife, Clara L. Holmes and C. J. Ferguson, setting forth that Charles T. and Clara L. Holmes of Charlotte in said County on, to wit, the 20th day of January. 1917. executed and delivered to the said William H. Tinsdale a mortgage deed of certain lands and premises in said Town of Charlotte with buildings and irrigation plant thereon described as follows:- "All and the same land and premises conveyed to us by Hattie S. Holabird by warranty deed dated September 26. 1917. recorded in vol. 20. at Page 20 of the land Records of the Town of Charlotte. Further described as being all and the same land and premises consisting of two hundred acres more or less. conveyed to the said Winooski Savings Bank. by deed of mortgage dated June 20. 1911. and recorded in vol. 21. at page 111 of said Land Records, to which deeds and the records thereof reference is here had. Said land and premises are further described as being the whole of the farm known and called the John Holmes farm, about one hundred acres of which the said John Holmes set to an orchard of apple trees which now exists on said farm. Together with all the buildings and erections of whatever name and nature now or

*Figure 6.3 Foreclosure document*

For family members with their heart and soul in the farm, the foreclosure language was stark. The decree ends with the stipulation that "all persons claiming or holding such property under them shall be foreclosed and forever barred from all equity of redemption in the premises." The result of the foreclosure action was that the Holmes family packed up and left for Middlebury, where they bought a house at 54 High Street at the foot of Chipman Hill. The property included two acres, the house and a barn. It was bordered to the north and east by publicly accessible Battell Park Land, a gift to the town of Middlebury by Joseph Battell.

It is unclear from available documents the true financial condition of the family at the time they left the farm. There were not able to make the mortgage payments, but did they bring assets with them?

The High Street house, still in the family, was a substantial, attractive structure. Middlebury town records indicate that Robert and Rena Holmes acquired the property on June 1, 1923 from Aurelius and Ida Sykes of Bristol for "one dollar and valuable considerations." We do not know what constituted valuable considerations. Perhaps the Sykes received some of the animals from the farm, such as horses or cows. In foreclosing on the farm, there is no indication that the stock stayed with the farm, so that would been an asset that remained with the Holmes.

Soon, however, Robert and Rena made another real estate move. Three months later, on Sept. 1, 1923, they sold the High Street house for $3,000 to C.D. Ordway of Burlington. The new deed read as follows:

> Shall and truly pay or cause to be paid to the said C.D. Ordway 11 certain notes in writing bearing date Middlebury Vermont September 1, 1923 and given for value received with interest annually at 5%.

Essentially, Robert and Rena turned around and mortgaged the property, with an immediate cash influx of $3,000 and 11 payments to follow. The document stipulated that "Holmes is to keep the buildings on said premises fully and safely insured for the benefit of this mortgage." This financial strategy—mortgage a property in order to live on it and regain ownership at a later date— was what C.T. had done with the farm, ultimately with an unfortunate outcome.

Although the scheduled payments ran through the years of the Great Depression, the family eventually came into full ownership of the property. It was not an easy path, however. Rena told me stories of the skimping they did to make the payments and sustain their lives in Middlebury during the Depression. On a visit to Middlebury, she pointed out that it was not good to waste sheets of toilet paper.

Their farming skills came in handy too: Robert planted various crops behind the house; Rena canned vegetables and fruits; they kept and monitored a tight family budget. Most importantly, Robert had a job through the depression years with the Central Vermont Public Service Corporation.

By the time of the move to Middlebury, C.T. and Clara were aging after a lifetime of farm labor, and C.T. had a heart condition.

Similar to when the extended family lived under one roof in the early years of the farm, there was little doubt that the family of six would move together into the High Street house.

Below is a picture (Figure 6.4) of the transplants from Charlotte taken in 1928—Robert, Rena, C.T., and Clara, with Marion and John in front—five years after their arrival.

*Figure 6.4 Family at High Street*

C.T. died of heart disease two years later in 1930 at the age of 73. The local paper noted his death, with a headline that said, "Was for many years prominent in apple industry and conducted many notable experiments," and reported:

> Practically his whole life was spent in the care of his father's farm and in the raising of and training of fine trotting horses. From 1906 until he came to Middlebury in 1923, he was one of the leading

> orchardists of the state in the development of the Lake View orchard to which he gave his whole attention, making it one of the finest orchards of the state. His experiments in apple culture were watched with great interest by apple men in this section of New England, and many orchardists are now benefitting by his experience.[19]

C.T. was buried in the cemetery on the hill behind the Congregational Church in Charlotte. Since that time, several generations of the Holmes family have been buried in this plot, including Clara who died in 1943.

With the loss of the farm and the move to Middlebury, what was the family's state of mind in 1923? The reality was that they went from a thriving, widely-known orchard and farm with a family deeply involved in the life of the local community to foreclosure and retreat to Middlebury 25 miles away.

Did C.T. and Robert foresee this outcome? Was the last decade of the farm a gradual descent into anxiety about money and the fate of the farm? Or, like so many farmers everywhere, did they believe that—one way or the other—they would go on. Things worked out for 100 years, why not now?

As a great grandson of C.T., I was born 12 years after his death. I spent considerable time with Robert, Rena, and Marion from my early teens until they passed away in their eighties. Marion lived at High Street for the rest of her life, a total of 70 years. Along with my father, mother, and sister, I visited High Street and the family camp on nearby Lake Dunmore during holidays. I went with Robert, Rena and Marion on camping trips; I hiked with Marion; I spent vacations from boarding school with them while my parents lived overseas. I look back now to remember what I heard about the loss of the farm and move to Middlebury.

What I realize is that the loss of the farm was an unspoken part of family history. Marion kept old picture albums of the farm and its people, which she shared on occasion. My uncle Charles Ross and my aunt Helen Patterson, children of Hannah and Jacob, talked about the prominent apple orchard and the highly successful horse business. But I heard little about my relatives who lived there. My grandparents and aunt took me to see the Holmes covered bridge just north of the property, but we never drove up Holmes Road to see the old property and the new residences.

In all of this, the story of the bankruptcy and what it meant to be pushed off the farm was never discussed.

I interpret the silence in two ways. First, the foreclosure and departure were a traumatic, sad and even embarrassing event for the family, but they dealt with the trauma with old-fashioned stoicism. As others have observed, this way of handling a hurtful event is not unusual among Vermonters. Indeed, the members of the family were not likely to fall apart emotionally over this turn of events. That's not who they were.

Second, the family focused quickly on what to do next. Where do we move? How do we take care of our older family members? How do we make a living? How do we shape the next chapter of family history? The members of the family were action-oriented, and this was their survival strategy. Rena and Robert realized that a move to Middlebury brought possibilities that did not exist on the farm and, from the start, they took advantage of what Middlebury had to offer.

## C. Explaining Failure: An Organizational Perspective

The Holmes farm was a business enterprise that encompassed four generations of adults (Nicholas, Jonathan, C.T., Robert) over a century. The fifth generation was represented by John, who was 7 years old when they left the farm. Since Vermont's earliest days, it appears that most farms were sustained beyond a single generation. We don't know from available data, however, how many farms survived for multiple generations and for how long.

Despite the lack of data on the early history of Vermont farms, there are many examples of family farms that were able to adapt successfully to changing conditions and sustain themselves. The Dakin farm in North Ferrisburgh, just a few miles south of the old Holmes farm, is a survival story. The farm was founded in 1792 and thrives today by selling Vermont products at two locations and through a highly successful mail order business. The Robb farm in Brattleboro was founded in 1907 and 105 years later left the milk business—milked their last Holstein—and turned to selling maple products and naturally grown meats.

In recent years, the farm-to-table movement and turning to organic products have helped sustain many enterprises. Also, the Vermont Land Trust has worked with farms to conserve land and address environmental issues, which reinforces the state's "brand" as a progressive, innovative place to live and farm. Numerous farms have successfully marketed themselves as environmentally progressive.

Yet, low milk prices continue to undermine the financial stability of Vermont dairies. Many decide not to reorient their business toward other products or are not in a financial position to do so, and they sell out. Generational factors play a role too. A successful 3,000-acre farm in Addison went on the market in 2019 and the Dubois brothers, all in their 60s and 70s, are tired of farming and want to "enjoy the rest of their lives." Their offspring are not going to take over.[20]

It is evident that the survival rate of multigenerational farms is influenced by both external and internal factors. Research reports, media stories, and word-of-mouth handed down over the years show that external factors have a profound impact on farm sustainability. In the Holmes case, World War I, the Spanish flu, and a volatile economy made survival much more difficult.

In addition, we need to look at factors within the family to assess the rise and fall of the Holmes farm. Specifically, how did characteristics and dynamics internal to the family affect the outcome? The scholarship of Pramodita and Sanjay Sharma at the University of Vermont Grossman School of Business is useful in looking back at the Holmes farm. There is a growing literature on family businesses and the factors that predict their success, and the Sharmas are leading scholars of this field of study. Their most recent work, *Pioneering Family Firms' Sustainable Development Strategies*, provides a framework for understanding and assessing the Holmes farm, including its ultimate demise.[21]

The Sharmas and their co-authors include families' environmental and social impact goals as a common theme. For the purposes of this analysis, however, the Sharma's six success factors for family firms are directly pertinent.[22] These factors are:

- Commitment driven by core values
- Control to implement strategic decisions
- Continuity of the family business beyond the incumbent generations
- Purpose to achieve (sustainable development) goals
- Professionalism embedded in accountability of self and others that sparks innovation via development of unique capabilities and the appropriate organization design and structure
- Partnerships with multiple stakeholders (to jointly achieve sustainable development goals)

The Sharmas point out that it is important that "family leaders and the dominant coalition (decision makers) have control to make related strategic choices and investment decisions for their business(es)." They also elaborate on the importance of continuity.

> Long leadership tenures with generational overlaps ranging from one to three decades provide the time period that the family firm needs to build the unique capabilities required to simultaneously achieve financial, social and environmental outcomes and embed such values, commitments, traditions and networks across generations.[23]

They point out that "Research suggests that leaders of business families with an expectation to continue their business beyond their tenure invest more in research and development (R&D) activities."[24]

Although the Holmes farm existed in an earlier era—early 1800s to early 1900s—and was smaller in scale than the firms studied by the Sharmas and their colleagues, the Sharma standards apply.

By the standards of the Sharma's research, how do we evaluate the Holmes multi-generational farm? Did its demise result from practices inconsistent with these standards? Or, independent of the standards, did external factors simply overwhelm the ability of the farm to survive?

The evidence suggests that the Holmes farm scored high on factors identified by the Sharmas.

*Commitment driven by core values*

The Holmeses brought a shared ambition to Vermont in 1788. They sought land, and they aimed to create a successful farm, first in Monkton, then in Charlotte. The tight family structure (father; mother, children, nearby relatives) and their religious tradition (first the Quaker meetings in Ferrisburgh, then the Congregational Church in Charlotte) reinforced family unity around hard work, success and gaining recognition.

*Control to implement strategic decisions*

As pointed out earlier, two younger brothers—Jonathan and C.T.—took precedence over older brothers in assuming leadership of the farm and its various businesses. We are not sure what factors led to this situation, but the letters of the family do not indicate conflict or tension around the matter of who, ultimately, would be in charge. In fact, even during hard times, there appeared to be family-wide support and affection for Nicholas, Jonathan, and C.T., the recognized farm leaders over 101 years. Although C.T.'s brother, William, and his family of nine left the farm for Proctor, available evidence does not suggest this was a move motivated by conflict over control.

*Continuity of the family business beyond the incumbent generations*

The Holmes family sustained itself on the farm over multiple generations. When Jonathan was disabled by his fall into the lake and endured severe pain for the rest of his life, C.T., supported by his brother, stepped in without a problem. The two brothers were already experienced and competent in the work of the farm, including the apple orchard and the horse business.

*Purpose to achieve goals*

The family, personified particularly by C.T., was driven to create a successful farm, build a first-rate orchard, and raise and race Morgan horses of the highest quality. Their sense of purpose was ever-present and, indeed, had a public dimension. Other Vermonters knew about the pioneering apple orchard and the horse business, and the Holmeses were comfortable with high expectations.

*Professionalism embedded in accountability of self and others that sparks innovation via development of unique capabilities and the appropriate organization design and structure*

The development of the well-known apple orchard is a prime example of the Holmes family striving to advance professional practice in a business that was emerging in Vermont and nationally. C.T. constantly sought improved approaches and innovative techniques, drawing on apple experts in state government and at the University of Vermont. Along the way, he tested numerous approaches and made adjustments in irrigation, spraying and fertilizing. He was known as an expert and spoke frequently at professional meetings. Likewise, the breeding of Morgan horses was approached with the highest professional standard, exemplified by breeding Holmes horses to Morgan 50 and selling horses in national and international markets.

*Partnerships with multiple stakeholders*

The family's most important business-related relationship was with the Vermont State Horticultural Society and its apple-growing members. The Society did an excellent job of convening members and sharing and disseminating knowledge. It was a voluntary organization, however, not a "partnership." The farm's main business connections were through *selling* the farm's agricultural products, apples, bricks from its small foundry, and Morgan horses bred at the farm.

This analysis indicates that the Holmeses exhibited characteristics that predicted success in building and sustaining a multi-generational enterprise: they had shared values; they were positioned to make necessary strategic decisions; they remained unified through several generations; they had a strong disposition toward professionalism and a scientific approach; they collaborated with other entities.

The farm was a dynamic enterprise that had many successes, overcame many challenges, and, according to the Sharma criteria, did many of the right things. So, what happened? There are several possible explanations.

Perhaps the farm had too many revenue-generating businesses (farm products; bricks; apples; horses) and failed to concentrate on one or two with the greatest potential to make money over many years. We know, however, that diversification is insurance against the vagaries of the market and natural disasters. In these terms, the farm seemed well positioned to make adjustments in bad times and flexible enough to exploit new opportunities. Since they did many things well, doing too much does not explain the farm's demise.

Perhaps the family became complacent as it moved into a fifth generation, and the quality of the work and of leadership began to wane. There is no evidence whatsoever that commitment and leadership had run its course by the 1920s.

C.T., in his early 60s, remained an energetic, passionate and focused leader and took active measures to address problems, particularly ongoing financial pressures. His only son, Robert, was experienced and skilled in the ways of the farm and loved its activities, from haying to spraying the apples trees to raising Morgan horses. It is true, however, that Robert did not have the strong, driven personality of his father. He was a milder, more soft-spoken man. Robert was just 38 when the farm failed, so we will never know what kind of leadership he would have brought to the farm.

Perhaps family unity around farming and the farm life began to wane. The women of the farm were crucial to the success of the enterprise from the very beginning, but by the early 1900s several Holmes women sought education beyond high school and aspired to be teachers. Robert's wife, Rena, had a ten-year career as a teacher but returned to Charlotte to marry Robert and support the farm. If the farm had survived, there is every indication that she would been deeply committed to the farm and would have brought a strong personality and sharp intelligence to her role.

The fall of the farm cannot be attributed to doing too many things, or diminished commitment, or a gradual loss of enthusiasm for the farming life. Rather, two intersecting factors—one internal; one external—explain the farm's demise.

First, there is the bane of almost all farmers. They need cash now on the expectation that sufficient revenue with flow to pay off loans and mortgages. C.T. twice took on substantial mortgages in order to sustain the farm, with the expectation—the hope—that income would flow at a level sufficient to make payments. In essence, he gave up ownership of the farm during the life of the loan and gambled on its future revenue potential. Or, to use the colloquial phrase, he bet the farm. Committed to keep the farm in operation, he had no better choice.

Second, events beyond their control—World War I, the Spanish Flu, a volatile economy, and years of dry weather—compounded the risk of the mortgage executed in 1917 with payments due through 1926. The prices of agricultural products fluctuated widely. Apple harvests plunged. This was not the time to be raising and selling fine Morgan horses. As it turned out, the deciding factor was the poor performance of the apple orchard. The family was heavily invested in this business (irrigation machinery, fertilizers and chemicals, spraying equipment, labor), and the investment did not pay off in the years before 1920 and after. They could not keep up with the mortgage payments, and the loan was called in. Thus, the end of the Holmes farm.

An ancient definition of a tragedy is this: despite good intentions and doing the right thing, things go badly. This was the case of the Holmes farm. It was a tragic end to 101 years of family dedication and operating the farm at a high level of quality. At the end, forces beyond their control overwhelmed their ability to survive. For farmers throughout history, this is not a new story.

## D. Transformation in One Generation: Education as a Catalyst

The move of the Holmes, Johns and Ross families from England to Americas was a life-changing decision, a decision driven by economics, religion and perhaps family dynamics. Likewise, the move from southern New England to Vermont's Champlain Valley was a consequential decision, a decision to better themselves, spiced by an ingrained spirit of adventure. The harsh reality in 1923, however, was that the Holmeses were forced off the farm and required to move on. Forced on them by foreclosure, the resulting dispersion—the diaspora—of the family altered the trajectory of family history in a profound way.

When May 1923 arrived, they were decisive about creating a new life in the shire town of Middlebury. Still, as one might expect, the family carried a degree of nostalgia about life on the old farm. They kept a horse at High Street. Below (Figure 6.5) is John with his mother looking from porch.

*Figure 6.5 John and horse, High Street*

Another hint of nostalgia was their purchase of a small farm in Leicester, a few miles south of Middlebury. The farm was managed by another family. I remember Robert, my grandfather, taking me to see the farm when I was a boy. It was a run-down operation, with a barn, cows, and pasture land. As a city boy I was shaken by the size of the animals, the odors, and the big machines. After a few years, they sold the Leicester farm.

More than anyone, Marion, who was 11 years old when they moved, kept the history alive. She maintained albums and archives and saved family records over her lifetime. She visited old cemeteries and created family trees. Her passion about family history was passed along to me when I returned to Vermont in 1974.

Although the farm remained imprinted in the memory of those who had lived there, the reality was that a family that had farmed in Vermont for more than 140 years became something else. The Holmes diaspora after 1923 opened up different career options, different places to live, and different aspirations. These different avenues reflected similar trendlines of America in the 20th century, especially increased mobility, a more urban population, and a more global consciousness.

Focusing on the C.T. Holmes clan—the family that lived on the farm at the time of the foreclosure—what happened over the next few generations? Scholars of Vermont history have asked a similar question. A 1970 Vermont Historical Society publication observed, "It would be worthwhile to know more about the first settlers and representatives of Vermont towns, and how their descendants have scattered across the nation and distinguished themselves in various ways."[25]

There are several related questions. Did members of the Holmes family retain the ambition and goal-orientation of previous generations? Did they build on the emerging motivation to attain a good education? Did they inherit the will to succeed personally and professionally? And where did they go to achieve their goals? Or looking more pessimistically, did the failure of the farm lead to a descent into a loss of self-respect, an inability to create success in a new place, or even a condition of poverty?

It is important to look at the record. C.T. and Clara had four children: Hannah Elizabeth, Robert, Maude, and Mildred. Maude moved away to marry and live in Arizona and Mildred died at a young age. So, for the purposes of this examination, I will trace the trajectory of Hannah and Robert and their children. Hannah and Robert lived in Middlebury and produced a long line of descendants. How did these two family lines develop over the next century?

The transformation of the family—in one generation—from a farm-centered existence to something else is revealed by what we see in their educational ambition and attainment. What follows is a case study of one family's rapid turn to higher education as an avenue of opportunity.

## *Legacy of Hannah Holmes Ross*

After getting her degree from the University of Vermont, Hannah taught mathematics in out-of-state public schools and later in Middlebury. She met Jacob Ross, a medical student, while both were enrolled at the University of Vermont, and they were married at the farm.

Hannah and Jacob initially lived in Richmond, where Jacob set up his medical practice and Hannah taught school. They moved to Middlebury in 1918 where Jacob was one of two town doctors. They had five children: Katherine. Austin, Charles, Helen, and Ruth. Ruth died as a young girl during the Spanish flu epidemic. Hannah and her husband had college degrees, and their children internalized a similar aspiration.

| Colleges of Hannah's children | Colleges of Hannah's grandchildren |
|---|---|
| Radcliffe (Katherine) | Radcliffe<br>Kenyon<br>Harvard |
| U. of Vermont (Austin) | U. of Kentucky<br>U. of Kentucky<br>U. of Kentucky<br>Stetson<br>Vanderbilt |
| U. of Michigan (Charles) | U. of Vermont<br>U. of Michigan<br>U. of Vermont |
| Radcliffe (Helen) | Harvard<br>Middlebury<br>Radcliffe<br>Harvard |

This presentation does not include advanced degrees. Charlie, for example, received a law degree and MBA from the University of Michigan on the G.I. Bill. There are various graduate degrees among the grandchildren. Several grandchildren and great-grandchildren attended private secondary schools. Unquestionably, there was an expectation that family members would attend college. This expectation continued down through the generations.

Following military service in World War II, Austin became a successful businessman in Louisville, Kentucky, in the construction field. After a stint with Austin in construction, Charlie practiced law in Burlington and served in Vermont and federal government positions. Katherine taught mathematics at Middlebury High School while raising Charlie and Helen after the death of their parents. After her marriage, Katherine raised her children and taught at a private school in Virginia. Helen was active in local politics and the civil rights movement.

The next generation includes business people, a doctor, public servants, non-profit leaders, and educators. Hannah's offspring lived in Virginia outside Washington, western New York, Boston, and in Kentucky and Florida. The children of the four siblings have lived in many locations along the Atlantic coast and Midwest, with three moving to Vermont along the way.

In sum, Hannah and Jacob's offspring received first-rate educations, went into a wide range of professions, and lived away from Vermont for most of their lives.

Despite the far-flung lives of the Ross clan, the family retained a strong connection to Vermont.

Charlie returned from Washington, D.C., and later ran for the U.S. Senate in 1974, the race won by Patrick Leahy who, 47 years later, is still serving. A noteworthy part of the campaign was that Charlie's siblings—Helen, Austin and Katherine—and several of their children converged on Vermont that summer to help with the campaign. Also, Charlie reprised his agrarian roots after his return to Vermont when he and his wife, Charlotte, built a house in Hinesburg and created Taproot Farm to breed and train Morgan horses.

The Ross clan remains present at Long Point on Lake Champlain in Ferrisburgh, where they own several camps and visit every summer. As of 2021, several Ross descendants live in Vermont.

## *The Legacy of Robert Holmes*

After arriving in Middlebury with Rena, his parents, and his children, Robert eventually took a position with the Central Vermont Public Service Corporation, where he sold appliances over a 23-year career until he retired in 1950 at the age of 65. Rena and Robert's children were 11 and 8 years old, respectively, when they came to Middlebury and enrolled in public school.

Robert and Rena and their children became active, enthusiastic members of the Middlebury community. Rena and Robert were members and officers of the Congregational Church and made many lifelong friends among the church members. Rena was active in supporting the so-called Poor Farm near Middlebury for locals experiencing poverty, both during the dire years of the depression and after.

Marion and John acclimated easily to town life and dressed accordingly (Figure 6.6):

*Figure 6.6 John and Marion, High Street*

Like the Ross branch of the family, the Holmes's soon realized the pleasures of outdoor activities away from the all-encompassing life on the Charlotte farm. They hiked the Long Trail, which snakes along the summits of the Green Mountains, and built a camp on nearby Lake Dunmore. Skiing became a family passion, facilitated by the location of the High Street house on the west slope of Chipman Hill.

Robert took up skiing at age 45, a move that was described by Burlington Free Press in 1955: "At 70, Middlebury's Robert Holmes lives just for winter—and skiing." The article pointed out that 15 years earlier he had broken his leg "but the next winter he was back on his skis again."[26] A 1940 Free Press story reported on Robert's broken leg, as follows: "R.N. Holmes, one of the leading members of the Middlebury Winter Sports Club, fell while skiing at the Bread Loaf ski development Sunday and broke bones of his left leg below the knee. "[27]

Looking back, Robert wrote, "have skied every year. Now age 76. It's true I broke my leg, only mishap to date. People who sit by the fire on a nice sunny day, temperature around zero, do not know what they are missing." He went on to describe climbing Mt. Mansfield in the dead of winter, "four miles of the toll road before there was a lift."[28]

Below is a picture (Figure 6.7) of Robert descending the hill behind the High Street house.

*Figure 6.7 Robert skiing*

Robert was instrumental in building the first ski tow at what became the Middlebury College Bowl. Closer to home, the college and locals built a ski jump on Chipman Hill, which became the venue for the jumping event at the college's annual winter carnival. The event drew hundreds of spectators, with rows of cars along High Street in front of the house. The jump (Figure 6.8) looked west from Chipman Hill.

*Figure 6.8 Chipman Hill ski jump*

With the proximity of the jump to 54 High Street., John learned to jump on the hill and became one of the leading jumpers on the college circuit. In the spring of 1936, during his senior year at the college, John won a competition and pair of beautiful Austrian jumping skis, an achievement which was noted in a local newspaper (Figure 6.9).[29]

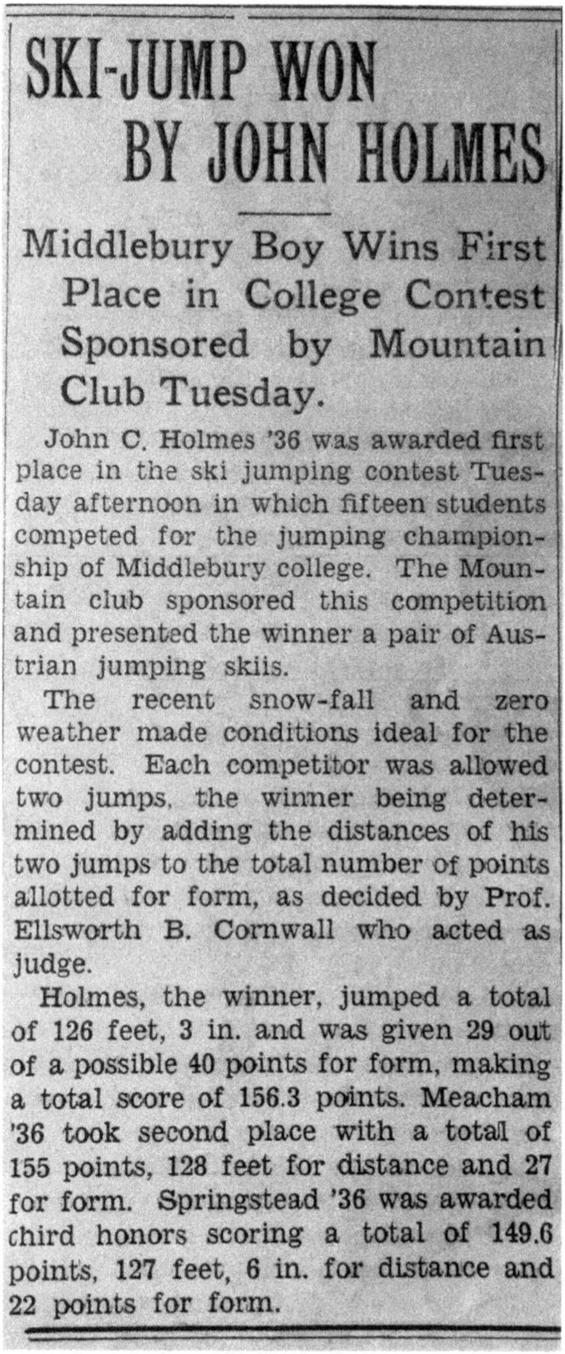

# SKI-JUMP WON BY JOHN HOLMES

## Middlebury Boy Wins First Place in College Contest Sponsored by Mountain Club Tuesday.

John C. Holmes '36 was awarded first place in the ski jumping contest Tuesday afternoon in which fifteen students competed for the jumping championship of Middlebury college. The Mountain club sponsored this competition and presented the winner a pair of Austrian jumping skiis.

The recent snow-fall and zero weather made conditions ideal for the contest. Each competitor was allowed two jumps, the winner being determined by adding the distances of his two jumps to the total number of points allotted for form, as decided by Prof. Ellsworth B. Cornwall who acted as judge.

Holmes, the winner, jumped a total of 126 feet, 3 in. and was given 29 out of a possible 40 points for form, making a total score of 156.3 points. Meacham '36 took second place with a total of 155 points, 128 feet for distance and 27 for form. Springstead '36 was awarded third honors scoring a total of 149.6 points, 127 feet, 6 in. for distance and 22 points for form.

*Figure 6.9 Clipping, John's ski jumping*

Similar to Hannah's line, Robert's family had a similar commitment to education beyond secondary school. There were fewer offspring along the way, so the branching is smaller.

Robert did not attend college; Rena, his wife, received a teaching degree and taught in schools in Vermont, New Hampshire and Indiana before marrying Robert.

| Colleges of Robert's children | Colleges of Robert's grandchildren |
|---|---|
| Middlebury (Marion) | (Marion never married) |
| Middlebury (John) | Middlebury<br>Middlebury |

Marion had no children. John married and had two children, who were educated in numerous secondary schools during John's career in the Foreign Service and later attended Middlebury College. His son (me) attained masters and doctoral degrees and has two children, with degrees, from Rollins College/Johns Hopkins University and Middlebury College/Vermont Law School, respectively. John's daughter has two children with degrees from Middlebury College and Duke University, respectively.

Marion was a longtime registrar of Middlebury College. John spent several years as a Special Agent of the FBI during World War II, then, after a short stint in the insurance business, entered the Foreign Service where he served in Germany, Japan, Washington, D.C., Thailand, and Vietnam. John's children have had careers mainly in the nonprofit world and have lived around the U.S. and overseas. Their children are in various professions at different locations in the U.S. His son returned to Vermont in 2017; his daughter and husband own the residence at 54 High St.

## *Overall Trajectory*

After the demise of the 101-year farm and their farming existence, Hannah Ross, Robert Holmes, and their families lived in Middlebury, and Middlebury became the launching point for the next generation. The family scattered widely to get excellent educations and pursue various professions. This trajectory—pursuing futures away from Vermont—was a radical departure from a past defined by farming in one place.

The Holmes diaspora and redirection occurred *in* ***one generation***.

## E. Fate of the Property

Owning land is satisfying and reassuringly tangible. Land is a source of income for those who use it for growing crops and raising animals. But there is a deeper psychological connection. The land has an enveloping presence, and it defines a part of who you are. To lose your land is to lose a piece of yourself. After the Holmeses left the farm, the land and buildings were gone for good.

Still, whether out of nostalgia or curiosity, the family kept an eye on the old property.

The property remained in foreclosure until the Thurber family bought it out of bankruptcy in 1936. The family revived the farm and raised pigs, cows, and sheep. To create pastureland, they removed the old apple trees. Sky Thurber, who grew up on the farm with his two sisters, recalled:

> They spent the early years clearing out the orchard. I remember a big bulldozer uprooting apple trees. They were pushed over the bank into what we called the "stump dump."[30]

The elimination of the trees was the final, symbolic closure of the Holmes era. The orchard had been a source of pride for the family, and now even the trees were gone. Among the apple trees that survived the dozer, here is a badly decayed tree (Figure 6.10) still standing in 2021.

*Figure 6.10 Decayed apple tree*

Eventually, the homestead Nicholas built with Holmes bricks was dismantled, and other buildings were altered or taken down. The race track was returned to pasture. The road up to the old farm property is called Holmes Road, as pictured below (Figure 6.11) on a wintry day looking southeast toward the location of the old race track.

*Figure 6.11 Holmes Road sign © Emily Anderson*

The beautiful Holmes covered bridge over Holmes Creek remains in place, an historical curiosity that is still the only route along the lake in that part of Charlotte.

While Holmes Bay, Holmes Creek, Holmes Creek Bridge, Holmes Road, and four surviving apple trees are still there, the Holmes family is long gone. Today, like so many old farmsteads in Vermont, the property has been transformed into a residential community with several up-scale houses across its 200 acres. To return to the property, one must rely on imagination—and old pictures—to reconstruct how it looked in 1900.

## *An Environmental Calamity Averted*

The property and its nearby shoreline might have looked very different today. In the late 1960s there was a serious threat that was thwarted by citizen action, including active involvement by members of the Holmes family. The Central Vermont Public Service Corporation, Robert's employer until 1950, announced that the Thurber family, owners of the property, had signed a purchase option of $260,000 in 1967 to sell 140 acres to the company for the purpose of building a nuclear plant.

The plan was for the Thurbers to keep the remaining farm acreage and continue to live there.[31] As a Thurber relative said later, "I don't think they had intentions of going anywhere. They were among the folks who believed that nuclear power was a safe and reliable option."[32] Their motivation to sell was driven by a financial bind. Property at that time was taxed according to its development potential, which was considerable, with annual assessments rising quickly. The Thurbers could no longer afford the taxes. It is relevant to note that this was before Vermont passed legislation to provide tax relief for owners who worked the land.

The land sale and the plan to build a nuclear plant on the property become known to Charlotte citizens at Town Meeting in 1968, and by October sparked a full-page advertisement paid for by the Lake Champlain Committee. The ad argued against the plant and outlined its perceived dangers. We know now that a nuclear plant would have been a disaster for the health of the lake, the beauty of the shoreline, and the culture of one of Vermont's beautiful small towns. The plant would have included a 400-foot smoke stack, which would be lit at night for aircraft safety, and a water exchange that would heat the lake and certainly kill aquatic life.

Like the nuclear plant in Vernon, it would have eventually been closed down, but the damage would have been done. A local newspaper transposed a nuclear tower on a picture of the shoreline (Figure 6.12), and one can envision the shocking intrusion on the land. The Holmes Covered Bridge is partially visible beyond the spit of land that reached into the lake to the north of the Holmes property.[33]

*Figure 6.12 Nuclear plant silo*

The proposed plant became a public *contretemps* in 1969, with Vermont politicians, including Sen. George Aiken (he was in favor of the project) and the federal Atomic Energy Commission becoming involved. Along the way, a survey of Charlotte citizens revealed that 60% opposed the plant, with 22% in favor. A public meeting on Sept. 11 in Patrick Gym at the University of Vermont drew 1,500 people and featured representatives of the AEC.[34]

Similar to other opponents, Charles Ross, C.T.'s grandson and a member of the Lake Champlain Committee, and Dan Kiley, a nationally acclaimed landscape architect who lived in Charlotte (and my future father-in-law) could see into the future. They were effective partners in fighting off the plant. In the end, the plant proposal was scrapped. Although no longer in the hands of the Holmes family, the land was saved from a disaster.

## F. Mobility and Tradition

Faced with diminished prospects, Vermonters have often looked outside the state to restart their lives. In the 1880s, when many Vermont farms were failing and western land became subsidized by federal legislation, there was a major out-migration. The era after the failure of the Holmes farm in 1923, however, was a different historical era. A few years after recovering from the dislocations of World War I, the Great Depression hit and with it came tales of the Dust Bowl and social upheaval. All this dampened enthusiasm of Vermonters for going to distant places.

After the foreclosure, the family—C.T.'s clan—went 25 miles down the road to Middlebury. In time, however, the Holmeses were shaped by societal changes and the emergence of the "modern age." The 30s and 40s brought world war, the emergence of the U.S. as an economic and political powerhouse, a true national transportation system, a growing sense that an advanced education was an avenue to vocational success, and good jobs that took people to cities. These forces influenced many Vermonters, including the Holmes offspring. The generation after Robert and Hannah looked elsewhere for opportunity. They grew up on the farm and in small-town Middlebury but the wider world beckoned.

As for the Holmeses and others who left Vermont, did they become different people? Did they lose their identity as Vermonters? Knowing that the social forces that touched the rest of the nation also touched Vermont, did those who stayed become different people?

With clues from the history of the Holmes family, the concluding chapter explores the Vermont character and reflects on what it means to be a Vermonter in 2021.

## Chapter Seven

# Reflections on the Vermont Character

*Vermonters are hard-working, resilient people who care about others and their community, have a sound moral foundation, work hard, know a lot, prefer things on a small scale, love the outdoors and the beauty of their state, and find humor in life's absurdities.*

THIS INQUIRY STARTED WITH THE STRAIGHT-FORWARD AIM of recapturing the 101-year history of the Holmes farm and orchard. I retained family archives collected by my aunt over 50 years, and then I added information about the orchard I found during my time in Vermont from 1974–1987. Since there are no previous case studies of Vermont farms from the era of the 1800s and early 1900s, I figured the project would be a useful addition to the literature on the history of Vermont agriculture.

All fine and good, but I soon realized that the farm history had personal dimension for me. In studying the photographs, the events, what people said and what they did, there was instant familiarity. Perhaps this is to be expected—I have the DNA of the people who lived on the farm.

An unexpected aspect of my inquiry was that "instant familiarity" went beyond my grandparents, father and aunt. I felt that I knew them all. Something about the way C.T. spoke and acted hit close to home. The same with Jonathan and Hannah. I identified with the entrepreneurial Nicholas, who decided to move the family to the shores of Lake Champlain. All of this opened a window on the family and on how I have been shaped by their history and attributes of character. It also brought forth insight into the Vermont character.

It is pertinent to point out that "character" has two general uses. First, it refers to the behaviors, attitudes and idiosyncrasies that characterize a person or people. Vermonters, it is claimed, are fiercely independent. This is a distinctive part of their collective character. Second, "character" also refers to the positive attributes that define a person or a people of "good character." These attributes include such things as honesty, generosity, truth-telling, civic mindedness, etc. Do Vermonters bring positive attributes such as these to their lives? In describing the essence of what it means to be a Vermonter, I address both aspects of character: (a) the idiosyncratic characteristics of Vermonters ("Don't look down your nose at me because you are rich") ; (b) the attributes that define what it means to be a good person ("Moose had a bad accident, and we need to help with the chores and bring supper.")

## A. Legacy of the Holmes Farm

What were the personal characteristics—the motivations, the ambitions, the attributes of character—that brought the Holmes family to Vermont in the 1700s, caused them to build the farm and orchard on Lake Champlain, and shaped the diaspora and what came after? These attributes stand out.

*Goal-orientation*

From the early 1800s, the family was driven to create a successful and highly regarded enterprise. C.T. aimed to develop a prominent orchard and have an impact on the science of growing apples. The ambition to excel seeped down through the generations and, eventually, this meant pursuing vocations away from Vermont: business, public service, medicine, law, and education. For many, it also meant excelling athletically, which became an avenue of success and recognition. Although Vermont offered many avenues of professional success, members of the family internalized a *national* standard for measuring how they were doing, and this led to a wider playing field.

*Affinity for action*

If nothing else, farm work imbues an affinity for action and getting things done. Members of the family sought outlets for their high energy and work ethic, and this led to trying new things (e.g., an apple orchard; raising and racing horses) and later going elsewhere to pursue opportunities. Physical labor was inherent in farm activities and, as such, they brought "sweat equity" to the enterprise. Their days were full, and they were high-energy doers.

*Aspiration for education*

Although schools of mixed quality served rural Vermont, the schools symbolized an important value. The family insisted that their children receive a basic education in the core subjects. Then, over the years, several Holmes and Ross women pursued teaching degrees and became educators. The move to Middlebury opened up college as a realistic possibility, and soon college became an expectation. For the family, a good education was the source of personal improvement and training in various vocations. To use a phrase from psychology's Carol Dweck, their passion for education reflected a pervading "growth mindset."

*Resilience and grit*

For more than 100 years, the Holmes family overcame the many challenges of running a farm: bad weather, illness and injury, managing multiple farm businesses, family dynamics, balancing the budget, etc. Rather than pausing and mourning the loss of the farm in 1923, the Holmeses and Rosses got busy in Middlebury with jobs and helping their children get a good education. They were resilient, gritty people.

*Expertise and knowledge*

Apart from a formal education, members of the family developed a deep reservoir of expertise and knowledge. The farm enterprise demanded that they be cognitively sharp. Otherwise, how could they have run a farm with a diverse set of needs, managed a successful brickyard, created a leading orchard, and raised, trained, and raced highly-rated horses? In the days before most people went off to college and intelligence was supposedly measured with an SAT test, they displayed a keen intelligence and knew a lot.

These characteristics are not the full picture, of course. There were other attributes common to the family: appreciation of the beauty of nature and the Vermont landscape, church-going (Quaker meeting house in Ferrisburgh, the Charlotte Congregational Church), love of play (games and competitive sports), and keeping in touch (letter-writing; family gatherings)

One other quality merits a comment. It has been said that one must look deeply to find a Quaker's sense of humor. From their Quaker heritage and from the hardships they faced, the Holmeses were a serious people who were devoted to farm life. They had an appetite—even passion—for play, but did they have the Vermonter's wry sense of humor? Because this kind of humor is subtle and low-key, the historical record is likely to miss it. I don't recall an undercurrent of humor from my grandparents who lived at the farm, but my father had a sly humor. After finishing a task, I pondered what he meant when he told me that I was "half-smart." He had a smile on his face, but was that a compliment or a subtle dig?

With memories of the farm still real among the grandchildren of C.T. and Clara (Marion, John, Katherine, Charlie, Austin, and Helen), the family converged on Charlotte in 1974 for a family reunion. Children, grandchildren, and members from other branches of the family attended. A group picture taken on the steps of the Charlotte Congregational Church included 63 family members.

This was a symbolic closing of the circle for the family and a once-in-a-lifetime event. Forty-seven years later, there has not been a similar gathering of the extended family. Like the migration that occurred after the foreclosure of the farm in 1923, the family scattered—again—to the winds.

## B. Being a Vermonter

The characteristics of the Holmes family—goal-oriented, ready for action, seeking education, resilient, and knowledgeable—describe, in part, what it means to be a Vermonter. I also looked at other sources of evidence: Vermonters observed in their native habitat, what others have written, and what I have seen and heard since my move home in 2017.

So, in the end, what do I know about what it means to be a Vermonter in 2021? Here are some thoughts about taking the measure of a people.

1. Since few people are perfectly consistent, the following attributes of character represent what one sees most of the time, but not necessarily all the time. Making thoughtful generalizations is what defining a people is all about.
2. Given the different dimensions of good character, it is useful to employ a conceptual framework in interpreting character. Drawing on the work of scholars who study character, I frame my generalizations below around four kinds of character: moral/ethical, performance, intellectual, and civic.[1]
3. Identifying attributes may create such a long list that the exercise obfuscates the *essence* of a people. To avoid this problem, I take a gulp and offer a single sentence that aims to capture the core identity of Vermonters.

I believe that Vermonters have demonstrated—and *still* demonstrate—these distinctive attributes of character:

*Moral/ethical character (integrity, truthfulness, honesty, and gratitude)*

A lapse in ethical behavior is very rare among Vermonters. Telling the truth is a core value. Vermonters are generous in expressing their gratitude to others and in saying thanks.

*Performance character (grit, perseverance, resilience, goal orientation, and work ethic)*

Vermonters often face challenging circumstances (cold weather, tight deadlines, and accidents) and are not complainers. They pride themselves in having high standards. They do not have a 9-to-5 psychology in their work. They bring "sweat equity" and grit to their lives and a degree of stoicism in the face of misfortune.

*Intellectual character (respect for facts and the truth, curiosity, and thirst for knowledge)*

Vermonters respect facts and the truth. They love to explore new ways of doing things. They know a lot. They seek education.

*Civic character (concern for others, concern for one's community, generosity, serving others)*

Vermonters want things on a small scale and desire face-to-face interaction. They take pride in where they live and are generous towards those who have troubles. They volunteer. They have an aversion to arrogance and self-importance. Family is important. The annual Town Meeting Day is important.

Vermont is sufficiently diverse that no single list captures it all. Beyond the above attributes, we see other things. Vermonters appreciate the beauty of nature. They have an affinity for simple things in life (boating, fishing, gardening, reading, and family time). There is an innate frugality that one finds in yard sales, fixing your own car or snowmobile, and growing your own vegetables.

Real Vermonters take delight in the mundane absurdities of life and have a sense of humor about it. This capacity is a sly, beneath-the-radar behavior that gets missed unless you have an ear for it and a sense of irony.

Nevertheless, knowing the pitfalls of generalizing about a people, I am comfortable in asserting the essence of what it means to be a Vermonter in a single sentence:

> Vermonters are hard-working, resilient people who care about others and their community, have a sound moral foundation, work hard, know a lot, prefer things on a small scale, love the outdoors and the beauty of their state, and find humor in life's absurdities.

The people who lived at the Holmes farm and other Vermonters I have observed throughout my life demonstrated these attributes, and I am shaped by them. It is part genetics and part knowing family and Vermonters. It's not a perfect match, of course, but it is who I think I am. It's good to be home.

## C. Always the Land

But time marches on. Vermont is not a monolithic society and faces challenges with the potential to tear it apart: an opioid epidemic; an undercurrent of prejudice among some Vermonters; too many Vermonters in poverty and lacking adequate health care; an epidemic of obesity with related health complications; environmental hazards such as toxic runoff into Lake Champlain and other waterways; urbanization of communities reinforced by affluent newcomers aiming to replicate an urban culture; a lucrative tourist industry that often produces a culture conflict between wealthy visitors and native Vermonters; too few decent-paying jobs for young people; too many over-built houses and landscapes.

This is the complexity of modern life. Yet, with all the forces at play in the modern age, let's remember that, in the beginning, acquiring and working the *land* was a driving motivation. In the 1700s, Vermont was *terra incognita* for most settlers coming from elsewhere in New England to take over a densely wooded property, usually sight unseen. Indigenous people, of course, had long occupied many sections of Vermont and lived off the land, but for the newcomers it was unoccupied land that promised a better future.

Simon Winchester in his book, *Land: How Hunger for Ownership Shaped the Modern World* (2021), examines how land and its possession tells us much about human history. He captures two opposed perspectives, the first the romantic notion that land should not be owned by anyone as expressed by Jean-Jacques Rousseau in 1755:

> The first person who, having enclosed a plot of land, and took into his head to say this is mine, and found people simple enough to believe him, was the true founder of civil society. What crimes, murders, what miseries and horrors would the human race be spared, had everyone pulled up the stakes or filled in the ditch and cried out fellow men: "Do not listen to this imposter. You are lost if you forget that the fruits of the earth belong to all and the earth to no one."

The realities of history have defied Rousseau's idea. Buying, cultivating, and exploiting land for one's benefit and satisfaction has prevailed in almost all places, and the Holmes/Johns/Ross clan were no different. They shared the sentiment of Anthony Trollope expressed in an 1867 novel quoted by Winchester: "It is a comfortable feeling to know that you stand on your own ground. Land is almost the only thing that can't fly away."[2]

Although Winchester demonstrates how mankind has damaged this precious gift, the Holmeses were deeply connected to the land, the landscape, and the transitions of the land over four distinct seasons. They built on it, cleared it, raised crops and animals on it, raised families on it, and marveled at the utter beauty of their surroundings.

Love of the land remains the one unalterable characteristic of Vermonters. Today, whether living in Burlington or Bristol or Rutland, Vermonters are connected to the land. It is a psychological connection ("The land is part of who I am.") and a physical connection (an admiring look at Camel's Hump; a hike on Chipman Hill; setting up camp at Button Bay State Park).

Even here, however, there are reasons for concern. Frank Bryan lamented, "The disjunction between what Vermont is and what Vermont looks like." Of the crisis Vermont faces, he wrote:

> Its solution cannot be reached unless we return again to the rooted values, the guts, of what it was that gave our forebears the courage to build Vermont in the first place. Among the values are a certain respect for, and fairness with, the culture of locality; an appreciation of the value of variety in human affairs—without which, indeed, there would be no art, no literature, no passion at all in the human soul; and most important, an understanding of the nexus between the land that sustains the people and the people who sustain the land.
>
> For there is one truth that must again be realized if Vermont, indeed, if the planet itself, is to saved: We are all farmers, from those who till the soil to those who hire others to do it for them.[3]

Bryan went on to prescribe a way to return to "rooted values":

> Anyone who believes you can understand Vermont without suffering (yes, suffering) through ten Aprils, end to end, doesn't know nature, can't understand that special relationship between the land and its people that stand behind Vermont character.[4]

Relying on Bryan's standard, I am doing okay. I experienced thirteen Vermont Aprils, from 1974–1987, in a small house set in the woods of East Charlotte. Since returning, I have lived four more Aprils along the windswept shores of Lake Champlain.

I know, too, that I inherited a love of place from the Holmes family who presided over the Charlotte farm and orchard for 101 years. So, at the end of all my exploring, I have arrived back where I started. I know Vermont and myself for the first time.

# Acknowledgements

A PROJECT OF THIS KIND—a story of a farm, a family, a state—is not a solo undertaking. I am grateful for the interest and help of the many people who expressed enthusiasm for the project and shared with me a love of Vermont and what it stands for.

Richard Watts, director of the Center for Research on Vermont, responded with immediate enthusiasm when I made a cold call about his potential interest in publishing my work. The Center is beginning a publications arm, and this book is among its first publications. The book was made possible by Richard, Emily Anderson, and Jessie Forand plus UVM students Noah Gilbert-Fuller, Martha Hrdy and Nick Kelm. Emily helped immensely in preparing the manuscript and inserting the pictures for publication. As a lifelong educator, I was thrilled to have students take a direct role in the project. Noah did useful research; Martha designed the cover; and Nick did first-level copy editing.

The work has benefitted greatly from the reading and comments of experts on Vermont and related topics, including David Moats, Nick Muller, Bill and Kate Schubart, Madeleine Kunin, Steve Terry, Roger Allbee, Jay Parini, Pramodita Sharma, and Travis Jacobs. Jenny Cole at the Charlotte Library and Dan Cole, President of the Charlotte Historical Society, guided me to valuable primary resources on the history of Charlotte. Sky Thurber, whose family bought the old farm in the 1930s, provided valuable insight into the farm's history. Nearby residents of the old farm, Harvey Allen and Jessie Bradley, led me to the burial ground on the property and the family headstones.

Sarah Sprayregen helped me contact current residents of the old farm property. Jay Bearman, a current resident of the property, extended his support for the project in various ways.

My wife, Toni, exuded constant enthusiasm for the project and repaired countless sentences and moments of faulty logic. She also tolerated my disappearance when I was immersed in the project tucked away in the study at home. Other members of the extended Holmes clan exuded enthusiasm and financial support for the project, even before seeing what I had said about the family. These relatives included the Holmes, Shattuck, Ross and Brooks branches.

Pete Ross took me on an informative tour of early Ross home sites in Huntington and, along with his brother, Chuck, supplied facts about the Ross family history. My cousin, Bill Brooks, Executive Director of the Henry Sheldon Museum of Vermont History in Middlebury, provided letters from the museum's collection between Hannah and Jacob Ross during World War I.

My late aunt, Marion Holmes, was the keeper of family history: pictures, letters, clippings, documents, and memories. Without her dedication to family history and her support of my efforts during the early stage of the project (1980s), the book would not have been possible. This book is dedicated to Marion.

# Notes

## Chapter One

1. Helene Stapinski, "Mobster Tale: A master of historical form turns his gaze on his family's past," Sunday Book Review, *The New York Times*, February 7, 2021, p. 17.
2. "What Is an American?" from *Letters from an American Farmer*, 1782, see *The American Tradition in Literature*, W.W. Norton, New York, 1962, 153–164; *The American Mind: An Interpretation of American Thought and Character Since the 1818's*, Henry Steele Commager, Yale University Press, 1950; *The Lonely Crowd: A Study of the Changing American Character*, Yale University Press, 1950; *Anti-Intellectualism in American Life*, Richard Hofstadter, Alfred A. Knopf, Inc. and Random House, Inc., 1962.
3. Graham Newell, "Latin Teacher and State Senator," *Vermonters: Oral Histories from Down Country to the Northeast Kingdom*, Ron Strickland, University Press of New England, 1998, p. 109.
4. Dorothy Canfield Fisher, *Vermont Tradition: The Biography of An Outlook on Life*, Little, Brown and Company, Boston, 1953, p. 6.
5. Ida Washington, "Dorothy Canfield Fisher's Vermont Tradition," *We Vermonters: Perspectives on the Past*, The Vermont Historical Society, Michael Sherman and Jennie Versteeg, editors, 1992, p. 169.
6. William Mares, "The Vermont State of Mind: Adventures in Schizophrenia," *We Vermonters*, 1992, p. 154.
7. Ibid., p. 57.
8. Gregory Sanford, "A Hardy Race: Forging the Vermont Identity," *We Vermonters*, 1992, p. 343.
9. *Toward a Revitalized Museum: The 2009 Interpretive Plan*, archives of the Sheldon Museum of Vermont History, 2009, pp. 12–13.
10. Dorothy Canfield Fisher, *Vermont Tradition: The Biography of An Outlook on Life*, Little, Brown and Company, Boston, 1953, p. 282.

## Chapter Two

1. Edwin Rozwenc, "Agriculture and Politics in the Vermont Tradition," *In the State of Nature: Readings in Vermont History*, H. Nicholas Muller and Samuel B. Hand, Vermont Historical Society, 1982, p. 154.
2. David Ludlum, *Social Ferment in Vermont, 1791–1850*, AMS Press, inc., 1966, pp. 8–9.
3. Quoted in *Ancestry of a Branch of the Holmes Family*, Feay Smith, Denver, Colorado, 1946, p. 2. In the introduction to his 98-page report on the Holmes ancestry, Smith indicated that "The original notes for this booklet were collected by Carlton Holmes prior to 1940. During that summer, he turned the notes over to me. Subsequently, I did further research and found a considerable amount of additional material."
4. Christopher S. Wren, *Those Turbulent Sons of Freedom: Ethan Allen's Green Mountain Boys and the American Revolution*, Simon & Schuster, 2018, pp. 2–3.
5. *History of Monkton Vermont, 1734–1961*, local history project, c1961 (undated), p. 32.
6. William Higbee, "The Friends' Cemetery in Monkton and Moving to the West," July 18, 1899, reprinted in *Around the Mountains*, The Charlotte Historical Society, 1991, p. 235.
7. Ibid., pp. 235–236.
8. Rowland Robinson, "Recollections of a Quaker Boy," *Atlantic Monthly*, Volume 88, p. 105 (quoted in "The Early Quaker Meetings of Vermont," Charles Hughes & H. Day Bradley, *Vermont History*, Volume XXIX, No. 3, July 1961, p. 158).
9. Quoted from Quaker records of Nine Partners, New York, by Feay Smith, *Ancestry of a Branch of the Holmes Family*, Feay Smith, Denver, Colorado, 1946, p. 2.
10. William Higbee, "The Friends' Cemetery in Monkton and Moving to the West," July 18, 1899, reprinted in *Around the Mountains*, The Charlotte Historical Society, 1991, p. 235.
11. Elise A. Guyette, *Discovering Black Vermont: African American Farmers in Hinesburgh, 1790–1890*, Vermont Historical Society, pp. 22–23.
12. Ibid., p. 189.
13. Ibid., p. 108.
14. *The Colonial Genealogist*, Volume V, No. 11, Summer 1972. p. 7.
15. Ibid., p. 11.
16. See "*Works of Historical Faith: Or Wrote Reason The Only Oracle of Man?*," Michael Bellesiles, p. 73, *Vermont History*, Spring 1989, Vol. 57, No. 2 (author refers to Zadock Thompson's biography of Ethan Allen, 1867).
17. Quoted by Abby Hemingway, *Vermont Historical Gazetteer*, , Volume 1, 1867.
18. Unattributed author, *Look Around Richmond, Bolton and Huntington, Vermont*, Chittenden Historical Society, Lillian Baker Carlyle, Editor, 1974, p. 43.
19. Gary Wills, *Lincoln at Gettysburg*, Simon & Schuster, New York, 1992, p. 65.
20. *Huntington Vt., 1786–1976*, community publication, Olga Halleck (ed), 1976, pp.1–3.
21. *Vermont History News*, May–June 1977., Volume 28, No. 3, p. 39.
22. Ralph Nading Hill, *Lake Champlain: Key to Liberty*, The Countryman Press, Taftsville, Vt., 1976, p. 4.

23. H. Nicholas Muller III, "The Tide Turns: Commodities on the Ebb: Immigrants on the Flow," *Lake Champlain: Reflections on the Past,* The Center for Research on Vermont, Burlington, Vt, 1989, P. 109.
24. Holmes family archives, news clipping, undated, unsourced.

## Chapter Three

1. William Higbee, "The Friends' Cemetery in Monkton and Moving to the West," July 18, 1899, reprinted in *Around the Mountains,* The Charlotte Historical Society, 1991, p. 237.
2. William A. Haviland and Marjorie W. Power, *The Original Vermonters: Native Habitants, Past and Present,* University Press of New England, Hanover, 1981, P. 8.
3. "A List of Desirable Farms and Summer Homes in Vermont," Report issued by Board of Agriculture, Vermont Agriculture Department, Montpelier, Vermont, 1897, p.8.
4. F.B. Beers Atlas of Chittenden County (1869), Charles E. Tuttle Company, Rutland, Vermont, 1971, p. 29.
5. *The History of Chittenden County,* D. Mason & Co., 1886, p. 536.
6. Undated (circa 1890), Holmes family archives.
7. *Burlington Free Press,* October 28, 1887.
8. The experiences of Sarah and Albert Williams on the island were researched and reported by Janice Garen, a cousin, who was filmed at the Charlotte Library on August 15, 2012, transcribed by Betty Ann Lockhart.
9. Charlotte Town records, Volume 3, p. 313.
10. "Memories of Mary Holmes Whalley as told by her daughter-in-law, Gwendolyn Whalley," *The American Descendants of Samuel Whalley of Holbeck England,* compiled by Harvey D. Allen, undated. p. 17. (Charlotte Town Library)
11. William A. Haviland and Marjorie W. Power, *The Original Vermonters: Native Habitants, Past and Present,* University Press of New England, Hanover, 1981, p. 1.
12. Ibid., p. 152.
13. Remarks by C.T. Holmes, "Managing 100 Acres of Large Trees," *Twelfth Annual Report, Vermont State Horticultural Society,* 1914, p. 83.
14. Edwin Rozwenc, "Agriculture and Politics in the Vermont Tradition," *In a State of Nature: Readings in Vermont History,* H. Nicholas Muller and Samuel B. Hand, Vermont Historical Society, 1982, p. 153.
15. David Donath, "Agriculture and the Good Society," *We Vermonters: Perspectives on the Past,* Vermont Historical Society, 1992, p. 218.
16. Holmes family archives.
17. Frank Bryan, "Political Life in a Rural Technopolity," *In a State of Nature: Readings in Vermont History,* H. Nicholas Muller and Samuel B. Hand, Vermont Historical Society, 1982, pp. 374–375. (Reprinted from Frank Bryan, *Yankee Politics in Rural Vermont,* The University Press of New England, 1974.)
18. Ibid., p. 375.
19. W.C. Funk, 'What the Farm Contributes Directly to the Farmer's Living," *Farmers' Bulletin,* U.S Department of Agriculture, #635, December 24, 1914. p. 4.

20. H.W. Hawthorne, "The Family Living from the Farm," U.S. Department of Agriculture, Department Bulletin No. 1338, August 1925, p. 1.
21. Ibid., p. 2.
22. Lillian H. Johnson and Marianne Muse, "Cash Contribution to the Family Income Made By Vermont Farm Homemakers, " Vermont Agricultural Experiment Station, Bulletin 355, Burlington, Vermont, June 1933, p. 1.
23. Jordan Kisner, "The Wages of Housework," *The New York Times Magazine*, February 21, 2021, p. 36.
24. Michael Foley, *Farming for the Long Haul: Resilience and the Lost Art of Agricultural Inventiveness*, Chelsea Green Publishing, White River Junction, Vermont, 2019, pp. 19–21.
25. Ibid., p. 335.
26. Ibid.
27. "Memories of Mary Holmes Whalley as told by her daughter-in-law, Gwendolyn Whalley," *The American Descendants of Samuel Whalley of Holbeck England*, compiled by Harvey D. Allen, undated. p. 17. (Charlotte Town Library)
28. Letters, Holmes family archives.
29. David Donath, "Agriculture and the Good Society," *We Vermonters: Perspectives on the Past,* Vermont Historical Society, 1992, p. 215–216.
30. Edwin Rozwenc, "Agriculture and Politics in the Vermont Tradition, " *In a State of Nature: Readings in Vermont History*, H. Nicholas Muller and Samuel B. Hand, Vermont Historical Society, 1982, p. 154.
31. The identifications of the Holmes sheep from the old picture were made in January 2021 by David Martin of the Vermont Sheep and Goat Association and Joseph Emenheiser of the University of Connecticut Extension Service.
32. *The History of Chittenden County*, D. Mason & Co., 1886, pp. 835–836.
33. *Business Directory of Chittenden County Vermont*, 1882–1883 (on fruit growers).
34. "Memories of Mary Holmes Whalley as told by her daughter-in-law, Gwendolyn Whalley," *The American Descendants of Samuel Whalley of Holbeck England*, compiled by Harvey D. Allen, undated. p. 17. (Charlotte Town Library)
35. *Look Around Hinesburg and Charlotte*, Chittenden County Historical Society, 1973, p. 37.
36. Murray Hoyt, *30 Miles for Ice Cream, The Stephen Greene Press*, Brattleboro, Vermont, 1974, p. 123.
37. Quoted in *Up in the Morning Early: Vermont Farm Families in the Thirties*, Scott Hastings and Elsie Hastings, University Press of New England, Hanover, NH, 1992, P. 38.
38. 1860 Census, Charlotte Grand List, Charlotte Town Hall.
39. *Charlotte News*, c1968, mimeograph, no author identified. (History of the Charlotte Meeting House).
40. Charles Ross, "Charles B. Ross: Morgan Horse Breeder," *Vermonters: Oral Histories from Down Country to the Northeast Kingdom*, Ron Strickland, University Press of New England, 1998, p. 141.
41. Ibid., p.141.
42. Ibid.
43. Ibid., p. 143.
44. Amanda Kay Gustin, "Joseph Battell and the Morgan Horse," *Vermont History*, March/April 2017, p. 70.
45. Ibid., pp. 70–77.

46. Holmes family archives.
47. "Charlotte Raceway," *Busy Work*, S. Russell Williams, 1980, Charlotte Historical Society, p. 25.
48. Holmes family archives.
49. "Sulky Racing Traditions Recalled by Robert Holmes," Quoted in *Addison County Independent*, May 1958.
50. *Official Gazette*, October 17, 1905, p. 1886.
51. *Walton's Vermont Register*, undated (c. 1905), p. 202.
52. Nicholas Muller III, "From Ferment to Fatigue? 1870–1900: A New Look at the Neglected Winter of Vermont," #7, Center for Research on Vermont Occasional Papers Series, Burlington: University of Vermont, 1984. (Reprinted in *Vermont Heritage: Essays on Green Mountain History, 1770–1920*, H. Nicholas Muller III and J. Kevin Graffagnino, The Center for Research on Vermont, 2020, p. 348.)
53. Ibid.
54. Kevin Graffagnino, "Arcadia in New England: Divergent Views of a Changing Vermont," *Vermont Heritage: Essays on Green Mountain History, 1770–1920*, H. Nicholas Muller III and J. Kevin Graffagnino, The Center for Research on Vermont, 2020, p. 375. (Reprinted from *Celebrating Vermont: Myths and Realities*, Nancy P. Graff, editor, Middlebury, 1991.)
55. Ibid.

## Chapter Four

1. Poem by A.J. Worden, news clipping, undated, unsourced, Holmes family archives.
2. Letter from C.T.'s teacher, dated March 8, 1831, Holmes family archives, photostat copy.
3. *History of Chittenden County*, W.S. Rann, editor, D. Mason & Co., Syracuse, NY, 1886, page unavailable.
4. *Ibid.*, p. 536.
5. Letters, Holmes family archives.
6. "Memories of Mary Holmes Whalley as told by her daughter-in-law, Gwendolyn Whalley," *The American Descendants of Samuel Whalley of Holbeck England*, compiled by Harvey D. Allen, undated. p. 17. (Charlotte Town Library)
7. Newspaper clipping, Port Henry, N.Y., undated, unsourced, Holmes family archives.
8. Letter from Charles Stowe from Santa Barbara, Cal. to Mary Whalley in Charlotte, March 28, 1920, Holmes family archive.
9. Harriet Beecher Stowe, *Uncle Tom's Cabin*, Young Folks Edition, M.A. Donahue and Co., Chicago, c1853.
10. *A Study of Education in Vermont*, Carnegie Foundation for the Advancement of Teaching, Montpelier, Vermont, 1914, p. 8.
11. Ibid., p. 11.
12. Ibid., p. 35.
13. Ibid., p. 42.
14. Ibid.
15. Ibid., p. 43.
16. Ibid., p. 26.

17. Wilma Farman, "Wilma Farman: One-Room School Teacher," *Vermonters: Oral Histories from Down Country to the Northeast Kingdom*, Ron Strickland, University Press of New England, 1998, p. 41.
18. Ibid., p. 43.
19. Ibid., p. 44.
20. Ibid., p. 45.
21. Holmes family archives.
22. George Dykuizen, *The Life and Mind of John Dewey*, Southern Illinois University Press, Carbondale, Ill, 1973, p. 25.
23. Chris Burns, "Between This Time and That Sweet Time of Grace: The Diary of Mandana White Goodnough," *Vermont History*, V.77, No.1, Winter–Spring 2009, p.40. (Quoted Lynn Y. Weiner, *From Working Girl to Working Mother: The Female Working Force in the United States*, 1820–1980, University of North Carolina Press, 1985)
24. Ibid.
25. Ibid., p.42.
26. Rena Johns letters, Holmes family archives.
27. Letter, Holmes family archives.
28. Holmes family archives.
29. Ibid.
30. Ibid.
31. Ibid.
32. Murray Hoyt, *30 Miles for Ice Cream, The Stephen Greene Press*, Brattleboro, Vermont, 1974. pp. 110–111.
33. Ibid., p. 126.
34. Ralph Nading Hill, *Contrary Country*, Rinehart and Company, New York, 1950, p. 142.
35. See the *Shelburne News*, Vol. 50, No. 15, April 15, 2021, p. 1., for background on the Shelburne Shipyard and quotes by Dan Sabick.
36. See Ralph Nading Hill's writings on the history of the steamboat era on Lake Champlain: *Contrary Country* (pp. 139–153) and *Lake Champlain: Key to Liberty* (pp. 235–258).
37. See *Ticonderoga* page on the Shelburne Museum website, 2021.
38. Holmes family archives.
39. See "The Calvin Coolidge Inauguration: Revisited: An Eyewitness Account by Congressman Porter H. Dale," *Vermont History*, Vol. 62, 1994, pp. 214–221.
40. "Report of the Second Annual Outdoor Meeting of the Vermont State Horticultural Society," *Ninth Annual Report, Vermont State Horticultural Society*, 1911, p. 111.
41. Mary Will's letters from Holmes family archives.

## Chapter Five

1. M.B. Cummings, *Nineteenth Annual Report*, Vermont State Horticultural Society, 1921, p. 37.
2. Julia Moskin, "A Shot of Elegance," *The New York Times*, February 34, 2021, p. D4.
3. Ibid., p. D5.
4. M.B. Cummings, *Nineteenth Annual Report*, Vermont State Horticultural Society, 1921, p. 37.

5. Victor I. Spear, "Vermont: A glimpse of its scenery and industry," State Board of Agriculture (Vt). Argus and Patriot, Montpelier, 1893, p. 13.
6. M.B. Cummings, *Twenty-Third Report of the Vermont State Horticultural Society*, 1925, p. 122
7. "Apple Culture in Vermont," Professor M.B. Cummings, *Vermont Agricultural Report*, 1911, p. 10.
8. Ibid., p.7.
9. M.B. Cummings, "Apple Culture in Vermont," *Thirteenth Annual Report*, Vermont State Horticultural Society, 1911, p. 25.
10. M.B. Cummings, *Nineteenth Annual Report*, Vermont State Horticultural Society, 1921, pp. 12–13.
11. B.P. Lutman, "Spraying Apples," *Ninth Annual Report*, Vermont State Horticultural Society, 1911, p. 113.
12. Ibid., 117.
13. Remarks by C.T. Holmes, "Managing 100 Acres of Large Trees," *Twelfth Annual Report, Vermont State Horticultural Society*, 1914, p. 82.
14. "High Pressure Orcharding in New England," Hollister Sage, *The Garden Magazine*, January 1912, p. 271.
15. Remarks by C.T. Holmes, "Managing 100 Acres of Large Trees," *Twelfth Annual Report, Vermont State Horticultural Society*, 1914, p. 83.
16. This document and cover letter to F.V. Bostwick on October 14, 1919 was provided by Susan Horsford and made available through the Charlotte Town Library.
17. Remarks by C.T. Holmes, "Managing 100 Acres of Large Trees," *Twelfth Annual Report, Vermont State Horticultural Society*, 1914, p. 83.
18. Ibid., p. 83.
19. Ibid.
20. see *Bennington Evening Banner*, October 18, 1910, and June 11, 1915; *Barre Daily Times*, November 7, 1910; *Middlebury Register*, December 22, 1911.
21. "High Pressure Orcharding in New England," Hollister Sage, *The Garden Magazine*, January 1912, pp. 270–271.
22. Remarks by C.T. Holmes, "My Experience with a Productive and Unproductive Orchard," *Seventh Annual Report, Vermont State Horticultural Society*, 1909, pp. 34–35.
23. Ibid., p. 35
24. Ibid.
25. Ibid.
26. Interview with Sky Thurber, Charlotte, Vermont, February 17, 2021.
27. "My Experience with a Productive and Unproductive Orchard", *Seventh Annual Report, Vermont State Horticultural Society,* 1909 p. 35–36.
28. "Report of the Second Annual Outdoor Meeting of the Vermont State Horticultural Society," *Ninth Annual Report, Vermont State Horticultural Society*, 1911, p. 111.
29. Ibid., pp. 111–112.
30. Ibid., p. 131.
31. Ibid., p. 115.
32. Ibid., pp. 135–136.
33. "Apple Culture in Vermont," Professor M.B. Cummings, *Vermont Agricultural Report*, 1911, p. 109.

34. Ibid., p. 113.
35. Ibid., p. 111.
36. Ibid., p.113.
37. *Vermont, An Apple Growing State*, Vermont Agriculture, Bulletin No. 11, October 1911. p. 10.
38. "High Pressure Orcharding in New England," Hollister Sage, *The Garden Magazine*, January 1912, p. 271.
39. "Managing 100 Acres of Large Trees," *Twelfth Annual Report, Vermont State Horticultural Society*, 1914, pp. 82–84.
40. Comment to author by Vermont orchardist, June 2021, anonymous.
41. Cameron Clifford, *Farms, Flatlanders and Fords: A Story of People and Place in Rural Vermont, 1890–2010*, The Clifford Archive, p. 15.
42. Looking Around Hinesburg and Charlotte, Chittenden County Historical Society, 1973, p.37.
43. *Burlington Free Press*, September 29, 1910.
44. Clipping, undated, *Burlington Free Press*, 1911.
45. Luther Putnam, follow-up comment to C.T. Holmes's presentation, "Managing 100 Acres of Large Trees," *Twelfth Annual Report, Vermont State Horticultural Society*, 1914, p. 85.

## Chapter Six

1. Interview by author with Helen Ross Patterson, July 18, 1985.
2. Isaac Fornarola, *Burlington Free Press*, April 18, 2021, p. 2A.
3. Chelsea Edgar, "Flu Seasoned: A central Vermont centenarian recalls another global pandemic," *Seven Days*, May 13–20, 2020, p. 45.
4. David Segal, "A Panic Barely Etched in Granite Memorials," *The New York Times*, May 17, 2020, p. 7.
5. Catherine Arnold (author of *Pandemic 1918: Eyewitness Accounts from the Greatest Medical Holocaust in Modern* History, St. Martin's Press, New York, New York, 2018*)*, quoted by David Segal, *The New York Times*, 2020, p. 7.
6. J.J. Ross, "How to Meet the Need of Rural Hospitals," *The Modern Hospital*, Vol. XIV, No. 2, February 1920, p.108 (read before the Second National Country Life Conference, Chicago, Ill. Nov. 8, 1919.)
7. Charles Ross, "Charles Ross: Morgan Horse Breeder," *Vermonters: Oral Histories from Down Country to the Northeast Kingdom*, Ron Strickland, University Press of New England, 1998, p. 142.
8. Ibid., pp. 141–142.
9. Chip Hart, "Ray Fisher: Hail to the Victor," *Green Boys of Summer: Vermonters in the Major Leagues, 1882–1993*, The New England Press, Shelburne, Vermont, 2000, pp. 94–95.
10. *Thirteenth Annual Report, Vermont State Horticultural Society*, 1915.
11. Stephen Terry, *Say We Won and Get Out: George D. Aiken and the Vietnam War*, Center for Research on Vermont, White River Press, Amherst, MA, 2020, pp. 151–152.
12. *Thirteenth Annual Report, Vermont State Horticultural Society*, 1915, p. 94.
13. "The Rise of Commercial Apple Orchards in Vermont," M.B. Cummings, *Nineteenth Annual report, Vermont State Horticultural Society*, 1921, p. 38.
14. Ibid., p. 45.

15. Ibid.
16. *The Apple Situation in New England*, published by the Connecticut and Maine Experiment Stations and the Extension Services of New Hampshire, Rhode Island and Massachusetts. c. 1926.
17. "Problems of the Vermont Apple Orchard," M.B. Cummings, *Twenty-Third Report of the Vermont State Horticultural Society*, 1925, p. 81.
18. *Twenty-Second Annual Report, Vermont State Horticultural Society*, 1924, p. 31.
19. Clipping, "special to the Herald," dated September 28, 1930, unsourced, Holmes family archives.
20. Molly Walsh, "Big Ag Sale: Is There a Market for a $23 Million Vermont Dairy Farm?" *Seven Days*, Nov. 20, 2019.
21. Pramodita Sharma and Sanjay Sharma, editors, *Pioneering Family Firms' Sustainable Development Strategies*, Edward Elgar Publishing, Northampton, MA, 2021.
22. Ibid., p. 6.
23. Ibid.
24. Ibid., pp. 6–7.
25. *News and Notes*, Vermont Historical Society, Volume 22, No. 3, November–December 1970
26. *Burlington Free Press*, April 13, 1955.
27. *Burlington Free Press*, Feb. 13, 1940.
28. Holmes family archives.
29. Holmes family archives.
30. Email from Sky Thurber, Feb. 6, 2021.
31. Mark Bushnell, "Vermont's other nuclear plant," Vermont Digger, Sept. 8, 2013, https://vtdigger.org/2013/09/08/bushnell-vermonts-other-nuclear-plant/.
32. Ibid.
33. "The road not taken: The semicentennial of Charlotte's nuclear power," *The Charlotte News*, Oct. 23, 2019, p. 1.
34. Ibid., p. 2.

## Chapter Seven

1. Since about 2015, Angela Duckworth, director of the Character Lab at the University of Pennsylvania, Richard Weissbourd, director of the Making Caring Common project at Harvard University, and other scholars have defined different dimensions of "character." These four categories emerged from this work.
2. Quoted by Simon Winchester in *Land: How Hunger for Ownership Shaped the Modern World*, Harper, 2021.
3. Frank Bryan, "We Are All farmers," *Vermont Odysseys*, Penguin Books, NY, 1991. p. 77.
4. Ibid., p. 8.

# Bibliography

Allen, Harvey D., *The American Descendants of Samuel Whalley of Holbeck England*, undated, Charlotte Town Library.

Arnold, Catherine, *Pandemic 1918: Eyewitness Accounts from the Greatest Medical Holocaust in Modern History*, St. Martin's Press, New York, New York, 2018.

Atkinson, Rick, *The British Are Coming: The War for America, Lexington to Princeton 1775–1777*, Henry Holt and Company, New York, 2019.

Baker, Lillian, editor, *Look Around Richmond, Bolton and Huntington, Vermont*, Chittenden Historical Society, 1974.

Bassett, Seymour T.D., editor, *Vermont: A Bibliography of it History*, The Committee for a New England Bibliography, G.K. Hall & Co, Boston, Massachusetts, 1981.

Beers. F.B., Atlas of Chittenden County (1869), Charles E. Tuttle Company, Rutland, Vermont, 1971.

Blanchard, Bruce C., Please Listen While I Think . . . I am a Vermonter, Four Corners Press, Wilton, CT., 1998.

Chittenden County Historical Society, *Look Around Hinesburg and Charlotte*, Vt, 1973.

Clifford, Cameron, *Farms, Flatlanders and Fords: A Story of People and Place in Rural Vermont, 1890–2010*, The Clifford Archive, 2011.

Commager, Henry Steele, *The American Mind: An Interpretation of American Thought and Character Since the 1818's*, Yale University Press, 1950.

Crockett, William Hill, *A History of Lake Champlain*, Burlington, Vermont, 1909 (new edition printed by McAuliffe Paper Co., c. 1936).

Cushing, John T and Arthur Stone, editors, *Vermont in the World War, 1917–1919*, by Act of the Legislature in 1919, Free Press Publishing Company, Burlington, Vermont, 1928.

de Crévecceur, St. Jean, *Letters from an American Farmer*, 1782, see The American Tradition in Literature, W.W. Norton, New York, 1962.

Donath, David, "Agriculture and the Good Society," *We Vermonters: Perspectives on the Past,* Vermont Historical Society, 1992.

Dykuizen, George, The Life and Mind of John Dewey, Southern Illinois University Press, Carbondale, Ill, 1973.

Fisher, Dorothy Canfield, *Vermont Tradition: The Biography of An Outlook on Life*, Little, Brown and Company, Boston, 1953.

Foley, Michael, *Farming for the Long Haul: Resilience and the Lost Art of Agricultural inventiveness*, Chelsea Green Publishing, White River Junction, Vermont, 2019.

Gilbert, C.L., editor, *Vermont Odysseys*, Penguin Books, New York, New York, 1991.

Graff, Nancy P., editor, *Celebrating Vermont: Myths and Realities*, , Middlebury College Museum of Art, Middlebury, 1991.

Graff, Nancy and Eric Borg, David Robinson, *At Home in Vermont: A Middlebury Album*, Rainbow Books, Middlebury, Vermont, 1977.

Guyette, Elise A., *Discovering Black Vermont: African American Farmers in Hinesburgh, 1790–1890*, Vermont Historical Society, 2010.

Halleck, Olga, editor, *Huntington Vt., 1786–1976,* community publication, 1976.

Hastings, Scott, *Up in the Morning Early: Vermont Farm Families in the Thirties*, University Press of New England, Hanover, NH, 1992.

Haviland, Wlliam A. and Marjorie W. Power, *The Original Vermonters: Native Habitants, Past and Present*, University Press of New England, Hanover, 1981.

Hemingway, *Abby, Vermont Historical Gazetteer*, Volume 1, 1867.

Higbee, William Wallace, *Around the Mountains: Historical Essays About Charlotte, Ferrisburgh and Monkton*, The Charlotte Historical Society and Academy Books, 1991.

Hill, Ralph Nading, *Contrary Country*, Rinehart and Company, New York, 1950.

Hill, Ralph Nading, *Lake Champlain: Key to Liberty*, The Countryman Press, Taftsville, Vt., 1976.

Hofstadter, Richard, *Anti-Intellectualism in American Life*, Alfred A. Knopf, Inc. and Random House, Inc., 1962.

Hoyt, Murray, *30 Miles for Ice Cream*, The Stephen Greene Press, Brattleboro, Vermont, 1974.

Ludlum, David, *Social Ferment in Vermont, 1791–1850*, AMS Press, inc., 1966.

Muller, H. Nicholas III and Samuel B. Hand, editors, *In the State of Nature: Readings in Vermont History*, Vermont Historical Society, 1982.

Muller, H. Nicholas III and J. Kevin Graffagnino, editors, *Vermont Heritage: Essays on Green Mountain History, 1770–1920*, The Center for Research on Vermont, 2020.

Rann, W.S., editor, *History of Chittenden County*, D. Mason & Co., Syracuse, NY, 1886.

Riesman, David and Nathan Glazer, Reuel Denney, *The Lonely Crowd: A Study of the Changing American Character*, Yale University Press, 1950.

Sharma, Pramodita and Sanjay Sharma, editors, *Pioneering Family Firms' Sustainable Development Strategies*, Edward Elgar Publishing, Northampton, MA, 2021.

Sherman, Michael and Jennie Versteeg, editors, *We Vermonters: Perspectives on the Past*, The Vermont Historical Society, 1992.

Simon, Tom, editor, *Green Boys of Summer: Vermonters in the Major Leagues, 1882–1993*, The New England Press, Shelburne, Vermont, 2000.

Smith, Feay, *Ancestry of a Branch of the Holmes Family*, Denver, Colorado, 1946.

Stowe, Harriet Beecher, *Uncle Tom's Cabin*, Young Folks Edition, M.A. Donahue and Co., Chicago, c1853.

Strickland, Ron, *Vermonters: Oral Histories from Down Country to the Northeast Kingdom*, University Press of New England, 1998.

Terry, Stephen, *Say We Won and Get Out: George D. Aiken and the Vietnam War*, Center for Research on Vermont and White River Press, Amherst, MA, 2020.

Versteeg, Jennie G., editor, *Lake Champlain: Reflections on Our Past*, The University of Vermont and the Vermont Historical Society, 1987.

Weiner, Lynn Y., *From Working Girl to Working Mother: The Female Working Force in the United States*, 1820–1980, University of North Carolina Press, 1985.

Williams, Russell S., "Charlotte Raceway," *Busy Work*, 1980, (collection of the Charlotte Historical Society).

Wills, Gary, *Lincoln at Gettysburg*, Simon & Schuster, New York, 1992.

Winchester, Simon, in *Land: How Hunger for Ownership Shaped the Modern World*, Harper, 2021.

Wren, Christopher S., *Those Turbulent Sons of Freedom: Ethan Allen's Green Mountain Boys and the American Revolution*, Simon & Schuster, 2018.

## Vermont State Publications

*Seventh Annual Report,* Vermont State Horticultural Society, 1909.

*Ninth Annual Report,* Vermont State Horticultural Society, 1911

*Twelfth Annual Report,* Vermont State Horticultural Society, 1914.

*Thirteenth Annual Report,* Vermont State Horticultural Society, 1915.

*Nineteenth Annual Report,* Vermont State Horticultural Society, 1921.

*Twenty-Second Annual Report,* Vermont State Horticultural Society, 1924.

*Twenty-Third Annual Report,* Vermont State Horticultural Society, 1925.

"A List of Desirable Farms and Summer Homes in Vermont," Report issued by Board of Agriculture, Vermont Agriculture Department, Montpelier, Vermont, 1897.

Carnegie Foundation for the Advancement of Teaching, *A Study of Education in Vermont*, Montpelier, Vermont, 1914.

*Vermont, An Apple Growing State*, Vermont Agriculture, Bulletin No. 11, October 1911.

Vermont Agricultural Experiment Station, Bulletin 355, Burlington, Vermont, June 1933.

*Vermont Agricultural Report*, 1911.

## Vermont History Publications

Center for Research on Vermont Occasional Papers Series

*History of Monkton Vermont, 1734–1961.*

Looking Around Hinesburg and Charlotte, Chittenden County Historical Society, 1973.

*News and Notes*, Vermont Historical Society

*The History of Chittenden County*, D. Mason & Co., 1886.

*Vermont History*

*Vermont History News*

## Newspapers

*Addison County Independent*

*Barre Daily Times*

*Bennington Evening Banner*

*Burlington Free Press*

*Middlebury Register*

*Seven Days*

*Shelburne News*

*The Charlotte News*

*The New York Times*

*Vermont Digger*

## Other Sources

*Business Directory of Chittenden County Vermont*, 1882–1883.

Charlotte Town records.

*Farmers' Bulletin*, U.S Department of Agriculture, #635, December 24, 1914.

Holmes family archives, Panton, Vermont.

Middlebury Town records

*Official Gazette*, October 17, 1905.

*The Apple Situation in New England*, published by the Connecticut and Maine Experiment Stations and the Extension Services of New Hampshire, Rhode Island and Massachusetts. c. 1926.

*The Colonial Genealogist.*

*The Garden Magazine*

*The Modern Hospital*, Vol. XIV, No. 2, February 1920.

*Toward a Revitalized Museum: The 2009 Interpretive Plan*, archives of the Sheldon Museum of Vermont History, 2009.

U.S. Department of Agriculture, Department Bulletin No. 1338, August 1925.

*Walton's Vermont Register.*

# Index

## F



## G



## H

## I



## J



## L



## M



## N



## O



## Q



## R



## S

# T



# U



# V



# W

CPSIA information can be obtained
at www.ICGtesting.com
Printed in the USA
FSHW010111171121
86195FS

9 781887 043946